India's Nuclear Fuel Cycle

**Unraveling the Impact of the
U.S.-India Nuclear Accord**

Synthesis Lectures on Nuclear Technology and Society

Editor
Paul Nelson, *Texas A&M University*
Taraknath V.K. Woddi, *Scientech: A Curtiss-Wright Flow Control Company*

India's Nuclear Fuel Cycle: Unraveling the Impact of the U.S.-India Nuclear Accord
Taraknath V.K. Woddi, William S. Charlton, and Paul Nelson
2009

© Springer Nature Switzerland AG 2022

Reprint of original edition © Morgan & Claypool 2009

India's Nuclear Fuel Cycle: Unraveling the Impact of the U.S.-India Nuclear Accord

Taraknath V.K. Woddi, William S. Charlton, and Paul Nelson

ISBN: 978-3-031-01361-4   paperback
ISBN: 978-3-031-02489-4   ebook

DOI 10.1007/978-3-031-02489-4

A Publication in the Springer series
SYNTHESIS LECTURES ON NUCLEAR TECHNOLOGY AND SOCIETY

Lecture #1
Series Editor: Paul Nelson, *Texas A&M University*
             Taraknath V.K. Woddi, *Scientech: A Curtiss-Wright Flow Control Company*

Series ISSN
Synthesis Lectures on Nuclear Technology and Society
ISSN pending.

India's Nuclear Fuel Cycle

Unraveling the Impact of the
U.S.-India Nuclear Accord

Taraknath V.K. Woddi
Scientech: A Curtiss-Wright Flow Control Company

William S. Charlton
Texas A&M University

Paul Nelson
Texas A&M University

SYNTHESIS LECTURES ON NUCLEAR TECHNOLOGY AND SOCIETY #1

ABSTRACT

An analysis of the current (February 2009) status and future potential of India's nuclear fuel cycle is presented in this book. Such a fuel cycle assessment is important, but relatively opaque because India regards various aspects of its nuclear fuel cycle as strategically sensitive. Any study therefore necessarily depends upon reverse calculations based on the information that is available, expert assessments, engineering judgment and anecdotal information. In this work every effort is made to provide transparency to these foundations, so that changes can be made in light of alternative expectations or subsequent information. This book should be of interest to policy experts, governmental specialists, technologists, nuclear technologists, and others seeking to understand and explain the associated facts and potential consequences of the recent U.S.-India civil nuclear accord.

KEYWORDS

US-India, nuclear cooperation, US-India nuclear cooperation, nuclear, fuel cycle, nuclear fuel cycle, thermal breeder, breeder reactor, India, nuclear history, thermal breeder reactor, India's nuclear history

Contents

Executive Summary

The civil nuclear accord recently adopted by the Governments of India and U.S. has heightened the necessity of assessing India's nuclear fuel cycle inclusive of nuclear materials and facilities. This agreement changes, for the case of India, the long-standing U.S. policy of attempting to prevent the spread of nuclear weapons by denying nuclear materials or technology transfer to states that have not adopted full-scope safeguards. The Pokhran II nuclear tests in 1998 have convinced the world community that India would never relinquish its nuclear arsenal, as would be corollary to adopting full-scope safeguards. This recognition has driven the desire to engage India through civilian nuclear cooperation. The cornerstone of any civilian nuclear technological support is separation of nuclear facilities into military and civilian facilities. In India's case, this separation is particularly complex, because its nuclear program was not designed with any separation in mind. Thus, for example, there is no significant physical difference between the "strategic" reactors India has reserved for possible military use and those designated as civilian under its separation plan. Rather the difference is completely administrative; "civil" reactors are those that will be under IAEA safeguards, and hence eligible to operate with internationally supplied fuel. To further complicate the issue, the Indian strategic reactors were designed to produce electricity, and the clear intention of India is to continue using them primarily for that purpose, as fuelled with indigenous uranium. Nonetheless, they have characteristics that make them candidates for production of weapon-grade plutonium, and India deems it essential to its doctrine of nuclear deterrence to maintain the possibility of using them for that purpose.

In this study, a complete nuclear fuel cycle assessment of India was performed to aid in assessing how entwined the military and civilian facilities in India are, and to identify potential issues to aid moving forward under the newly adopted agreement. To estimate the existing uranium reserves in India, a complete historical assessment of ore production, conversion, and processing capabilities was performed using open source information that suggests a sharp decline in indigenous Indian production of uranium occurred sometime in the late 1990's. Sensitivity of conclusions to this hypothesized decline was assessed, and comparisons to other independent results were made. Nuclear energy and plutonium production (reactor- and weapons-grade) were simulated using declared capacity factors and modern simulation tools. The Indian (Bhabha) three-stage nuclear power program entities and all the components of civilian and military significance were assembled into a flowsheet to allow for a macroscopic vision of the Indian fuel cycle. These assessments included historical analysis and future projections with various possibilities for use of resources. This detailed view of the nuclear fuel cycle suggests avenues for improvement in the system and emphasizes the necessity for technological collaboration. The fuel cycle that emerges from this study exploits domestic Indian thorium reserves with advanced international technology and optimization of its existing system.

To utilize any appreciable fraction of India's supply of thorium, nuclear breeding is necessary. The two known possibilities for production of more fissionable material in the reactor than is consumed as fuel are fast breeders or thermal breeders. This book analyzes a thermal breeder core concept involving the CANDU core design. The end-of-life fuel characteristics evolved from the designed fuel composition is proliferation resistant and economical in integrating this technology into the Indian nuclear fuel cycle. Furthermore, it is shown that the separation of the strategic and civilian components of the Indian fuel cycle can be facilitated through the implementation of such a system.

CHAPTER 1

Introduction

1.1 MOTIVATION FOR THIS STUDY

The Governments of India and the U.S. recently adopted a civil nuclear accord, with concurrence from the International Atomic Energy Authority (IAEA) and the Nuclear Suppliers Group (NSG). The intense debate surrounding this event, and extending over nearly three years, has highlighted the interest in an accurate assessment of India's nuclear fuel cycle. This agreement, which reverses three decades of U.S. nuclear policy toward India, expectedly created considerable controversy due to tensions between two widely held American foreign policy objectives: (1) strengthening bilateral relationships with emerging powers and (2) leveraging access to materials and technology for civil nuclear energy to prevent the proliferation of nuclear weapons. The U.S. has long sought to build a relationship with India, a rising power and ambitious nuclear state since it first exploded an atomic bomb in 1974. India reiterated its resolve to possess nuclear weapons with its second test in 1998. Since 1974, successive U.S. administrations pursued a policy of technology denial and nuclear trade isolation until India relinquished its nuclear arsenals. American policy laid important and long lasting impressions. Isolation has likely slowed down arms buildup in the sub-continent. However, no state can ignore the 32 years (1974-2006) of penalty that India has endured because of its decision to develop and possess nuclear weapons. But during this period, it has become clear that denuclearizing India was an unachievable objective even with a cohesive effort by the U.S., the IAEA, and the NSG.[1]

The Pokhran-II nuclear tests in 1998 convinced the U.S. that India would never formally and unilaterally restrict or relinquish its nuclear arsenals. This along with other diplomatic contacts during this same time frame drove a desire to work with India on a broader, strategic level through civilian nuclear cooperation. Within the broad nuclear non-proliferation regime, the cornerstone of any civilian nuclear technological support is the separation of military and civilian facilities. The degree of entanglement of the military and civilian facilities can only be assessed by a full-scale detailed view of the complete nuclear fuel cycle of India. Assessment of uranium reserves, plutonium production, energy generation, technology available and projects attempted facilitates the dual objectives of strengthening bilateral ties and preventing the spread of nuclear weapons through stronger cooperation.

1.2 OBJECTIVES OF THE STUDY

A great deal of speculation has occurred with regards to the Indian fuel cycle (both military and civil) since the initial agreement for nuclear cooperation between the U.S. and India was reached on

[1] M.A. Levi and C.D. Ferguson, "U.S.–India Nuclear Cooperation, A Strategy for Moving Forward," Council on Foreign Relations, CS No. 16 (June 2006).

July 18, 2005. Much of this speculation seems based on a lack of complete understanding regarding the technical details of the Indian fuel cycle and Indian nuclear facilities; however, some speculation is also a product of uncertainties in the status and disposition of various Indian facilities. The overall objective of the work presented here was to analyze and document the Indian fuel cycle especially with respect to its relevance to the U.S.-Indian Nuclear Cooperation Agreement of 2005. This book is intended to provide an unbiased and complete resource to aid policy-makers and decision-makers, both in regard to that agreement as well as in future cooperation with India. This book may also aid in public awareness of the status of the Indian fuel cycle. The specific goals of this work were as follows:

- Provide a complete description of all Indian nuclear facilities (including decommissioned facilities, currently operating facilities, facilities under construction, and facilities planned).

- Provide an historical assessment of the Indian fuel cycle and material production in India since inception to the present day (specifically December 31, 2006). This includes a determination of the present reserves of all nuclear materials in India.

- Provide a detailed technical analysis of the future status of the Indian fuel cycle and material production if the U.S.-India Cooperation Agreement had **NOT** been placed in effect.

- Provide a detailed technical analysis of the future status of the Indian fuel cycle and material production if the U.S.-India Cooperation Agreement continues on course.

This work is focused on the technical assessments for the Indian fuel cycle based on open source information on the Indian nuclear facilities and the usage of those facilities. Assumptions and uncertainties included in any of the models used here are explicitly declared. The basis for the somewhat unusual hypothesized sharp decline in Indian uranium production in the late 1990's is discussed in some detail.

CHAPTER 2

A Brief History of the Indian Nuclear Program

In this chapter, a brief developmental history of both the nuclear power and nuclear weapons programs of India is given. The nuclear history was derived from numerous sources and an effort was taken to try to supply the most accurate details possible. This historical reporting is provided to increase the awareness of the reader to the complexity of the Indian program, as well as to set the stage for the technical assessments of the Indian fuel cycle.[1] Assumptions and inferences derived have been explicitly stated whenever possible.

2.1 THE INCEPTION PHASE OF THE PROGRAM: 1944-1960

The beginning of the Indian nuclear program occurred prior to the independence of India from British rule. In 1943, Dr. Homi Jehangir Bhabha submitted a letter to the Sir Dorab Tata Trust to found a nuclear research institute.[2] The approval of this proposal in April 1945 led to the creation of the Tata Institute of Fundamental Research (TIFR). TIFR began operations in Bangalore in June 1945 with Bhabha serving as the first director.[3] In December 1945, Bhabha moved TIFR to Bombay and its official inauguration was on December 19, 1945.[4]

India became an independent state on August 15, 1947. A year after that, the government of India passed the Atomic Energy Act of 1948 leading to the establishment of the Indian Atomic Energy Commission (AEC).[5] The AEC would pursue in-depth studies on nuclear energy. The AEC consisted of three members: Dr. Bhabha, Dr. K.S. Krishnan, and Dr. S.S. Bhatnagar. [6]

At a press conference in Madras (presently called Chennai), Prime Minister Nehru spoke about the values of developing atomic energy, as follows "We are interested in atomic energy for social purposes. Atomic energy represents a tremendous power. If this power can be utilized as we use hydroelectric power, it will be a tremendous boon to mankind, because it is likely to be more available and cheaper than the building of huge hydroelectric works. Therefore, we are interested in

---

[1]There are a number of online resources available that give detailed compilations of the history of India's nuclear weapons program and the reader should consult these for additional information. Two excellent sources are the Nuclear Threat Initiative webpage (http://www.nti.org/e_research/profiles/India/Nuclear/2296.html) and the Nuclear Weapons Archive webpage (http://nuclearweaponarchive.org/India/IndiaOrigin.html).

[2]"History of TIFR," http://www.tifr.res.in/scripts/content_r.php?schoolid=&terminalnodeid=1100&deptid=.

[3]G. Venkataraman, Bhabha and His Magnificent Obsessions, Universities Press India, Hyderabad, p. 114 (1994).

[4]P. Balaram, "Icons of Industry and Philanthropy: The Tata Centenaries," Current Science, Vol. 86, No. 8, pp. 1051-1052 (April 2004).

[5] "Heritage of BARC," http://www.barc.ernet.in/webpages/about/hjb.htm.

[6]I. Abraham, The Making of the Indian Atomic Bomb, Zed Brooks Publishing, London, p. 61 (1998).

the development from the social point of view." A four-year plan was unveiled to develop India's nuclear infrastructure for nuclear material exploration and the application of nuclear energy in medicine. During the period, Dr. Bhabha began seeking technical information on reactor theory, design, and technology from the U.S., Canada, and the U.K. while negotiating the sale or trade of raw materials such as monazite and beryllium-containing ore.[7]

In August 1950, Indian Rare Earths Limited was established for recovering minerals and the processing of rare-earths compounds and thorium-uranium concentrates. Later in April 1951, uranium deposits were discovered at Jaduguda, in the state of Bihar and drilling operations commenced in December 1951.[8] The Jaduguda mine was the main source of uranium for the entire Indian nuclear program until the present day.

In 1954, significant changes occurred which led to a definite path for establishing a nuclear-weapons capability. On January 3, 1954, the Atomic Energy Establishment at Trombay (AEET) was created by the AEC. AEET led research on nuclear weapons technology and has been referred to as the "Indian Los Alamos." [9] The AEET was formally inaugurated on January 20, 1957 and was followed by the creation of the Department of Atomic Energy (DAE) on August 3, 1954 with Dr. Bhabha as Secretary.[10] The DAE was not under the regular control of the cabinet but reported directly to the Prime Minister. On January 12, 1967 in tribute to Dr. Bhabha, who died in an airplane crash on January 24, 1966, the AEET was renamed as the Bhabha Atomic Research Center (BARC). During his tenure, Dr. Bhabha also transferred all scientific initiatives from TIFR to AEET.

On May 10, 1954, Prime Minister Nehru sharply reacted to President Eisenhower's Atoms for Peace plan. He urged the Indian parliament to support plans to expand India's nuclear energy programs. He also declared that "the use of atomic energy for peaceful purposes is far more important for a country like India, whose power resources are limited, than for a country like France, an industrially advanced country."[11] Bhabha and Bhatnagar then conducted a series of meetings with British officials to request assistance in constructing a nuclear reactor and in converting uranium ores into metal for fabrication. He requested five tonnes of heavy water[12] for use in a planned Indian

---

[7]S. Bhatia, India's Nuclear Bomb, Vikas Publishing, Ghaziabad, p. 90 (1979).

[8]http://www.barc.ernet.in.

[9]The AEET would be later renamed the Bhabha Atomic Research Center (BARC) by Prime Minister Indira Gandhi on January 12, 1967 in tribute to Dr. Bhabha who died in an airplane crash on January 24, 1966.

[10]R.S. Anderson, "Building Scientific Institutions in India: Saha and Bhabha," Occasional Paper No. 11, Centre for Developing-Area Studies, McGill University, Montreal, Canada, p. 40 (1975).

[11]"Control of Nuclear Energy," Statement in the Lok Sabha (lower house of parliament) made on May 10, 1954. Available at http://www.indianembassy.org/policy/Disarmament/disarm3.htm.

[12]Heavy water (also referred to as D_2O) is a form of water that consists of two deuterium atoms (H-2 or D) bound to an oxygen atom. Deuterium is a heavy form of hydrogen consisting of one neutron and one proton [whereas regular hydrogen (H-1) consists of only one proton]. Heavy water (as opposed to light water or H_2O) exists in only trace quantities in ordinary water (about 1 part in 5000) and thus must be produced using isotopic enrichment techniques for use in reactor systems. Heavy water can serve as an excellent neutron moderator in reactor systems because it has an exceptionally low affinity for absorbing neutrons. This allows reactors moderated with heavy water to employ natural uranium fuel (as opposed to reactors using light water which must use enriched uranium fuel). To produce heavy water, ordinary water can be electrolyzed to make oxygen and hydrogen containing normal hydrogen gas and deuterium. The hydrogen can then be liquefied and distilled to separate the two species. Finally, the resulting deuterium is reacted with oxygen to form heavy water. No nuclear transformations occur in the production of heavy water.

research reactor. The British encouraged Bhabha to approach the Atomic Energy of Canada Limited (AECL) for the supply, as the U.K. had a deficit of heavy water available for domestic use.

On November 26, 1954, Prime Minister Nehru stated at the conference on "Development of Atomic Energy for Peaceful Purposes" that "atomic energy is a tremendous tool for the benefit of humanity, whether it is disease or poverty. It therefore becomes necessary for us to try not to lag behind in this, although we may not have the great resources that some other countries have." At the same conference, Bhabha presented the three-stage nuclear energy plan for national development (shown in Figure 2.1).

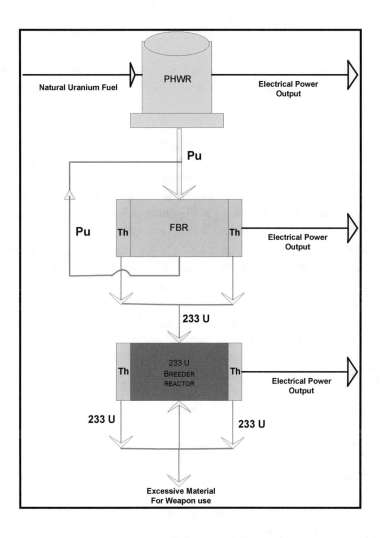

Figure 2.1: India's three-stage nuclear power production plan.

Under this plan, India would start its nuclear energy generation in the first-stage with natural uranium-fueled, heavy-water moderated Pressurized Heavy Water Reactors (PHWRs) to produce power and plutonium. The first-stage reactors would be based on the CANDU technology and be built with Canadian assistance. These reactors were planned to generate 420 GWe-yrs of electricity in its lifetime.[13] In the second-stage, plutonium separated from the spent fuel of the first-stage reactors would be used to power Fast Breeder Reactors (FBRs). The FBRs would then generate an additional 54,000 GWe-yrs of electricity. While generating power, the FBR's would also irradiate thorium in the blanket of the FBRs to breed U-233. In the third-stage, the U-233 bred from the second-stage would serve as fuel for the U-233 breeder reactors.[14] These U-233 breeder reactors would further generate 358,000 GWe-yrs of electricity and breed additional fissile materials. Breeder reactors are an advanced reactor design in which the reactor breeds additional fuel, during operation to produce electricity. Effectively, a breeder reactor produces more fuel during operation than it consumes. The breeder reactor achieves this through a design that highly conserves neutrons in the system. These reactors would produce enough material to fuel themselves and produce additional fissile material (plutonium and/or U-233), which conceivably could be used either for weapons or to fuel power reactors.[15]

The basis of three-stage-program was the indigenously available technology for production of natural uranium fuel assemblies, the vast reserves of thorium in India, and obtaining mastery of heavy water production and spent fuel reprocessing technology. When this program was devised, India did not have any existing power reactors and there were no commercial FBR systems anywhere in the world. The Indian government formally adopted this three-stage plan in 1958, thereby recognizing the importance of nuclear power as a sustainable energy source.

On the heels of Bhabha's nuclear power strategy, the atomic energy program in India grew rapidly. The budget for research in atomic energy grew by a factor of twelve from 1954 to 1956.[16] By 1959, the DAE consumed almost one third of India's research budget and the AEET employed over 1,000 scientists and engineers.

In 1955, the 1 megawatt thermal power (MW_{th}) APSARA research reactor was built with British assistance. APSARA went critical on August 4, 1956. APSARA was a light-water cooled and moderated swimming pool-type research reactor. It uses low-enriched uranium, plate-type fuel. The British-origin fuel for the reactor is safeguarded as per the supply contract but the reactor itself is not under IAEA safeguards. ASPARA was the first operating reactor in Asia outside of the Soviet Union though only days ahead of Japan's first reactor.[17]

During most of 1955, India was actively negotiating with the U.S. and Canada to acquire a research reactor with plutonium production capability. These negotiations began when members of

[13] A. Gopalakrishnan, "Evolution of the Indian Nuclear Power Program," *Annual Review of Energy and the Environment,* Vol. 27, p. 372 (2002).

[14] U-233 is a non-naturally occurring isotope of uranium that can serve as an excellent nuclear fuel (similar in many ways to U-235). U-233 is produced by irradiating Th-232, which produces Th-233, which then undergoes two subsequent b- decays to U-233.

[15] R. Chengappa, *Weapons of Peace: Secret Story of India's Quest to be a Nuclear Power,* Harper-Collins Publishers, India (2000).

[16] http://nuclearweaponarchive.org/India/IndiaOrigin.html.

[17] "Research Reactors," *Nuclear Review,* p. 17 (April 1996); and http://www.barc.ernet.in.

the U.S. Joint Committee on Atomic Energy visited India in early 1955 to promote the expansion of the peaceful applications of atomic energy under the Atoms for Peace Program. In September 1955, Canada agreed to supply India with a 40 MW_{th} research reactor. Canada also offered to pay all the foreign exchange costs of building the reactor. No strict safeguards on the use of the plutonium produced by the reactor were made other than the commitment by India, via an annex to the agreement, that the reactor and fissile materials it produced would be used only for peaceful purposes. The reactor was named the Canada-India Reactor (CIR).

On March 16, 1956, a contract was signed by the U.S. and India for supply of 18.9 tonnes of heavy water for the reactor. This agreement was based on the possibility that India's own Nangal heavy water plant may fail to operate sufficiently.[18] The agreement stated that "the heavy water sold hereunder shall be for use only in India by the government in connection with research and the use of atomic energy for peaceful purposes...."[19] Following this, the reactor was dubbed the Canada-India Reactor, United States (or CIRUS). [20]

The acquisition of CIRUS was a landmark event for both atomic energy developments in India and nuclear proliferation. Although the agreement required that the reactor only be used for peaceful purposes, it occurred before any international policy regulations were in place for nuclear technology transfers and consequently no provision for inspections were made. India refused to accept fuel from Canada and carefully avoided effective regulation for the reactor. India established a program to indigenously manufacture the natural uranium fuel for CIRUS so as to keep complete control of the produced plutonium. This fuel manufacturing program, led by metallurgists eventually succeeded in developing techniques for producing the precise high-purity material demanded by the reactor. On February 19, 1960, ten fuel elements for the first load of CIRUS at Trombay were fabricated.

CIRUS achieved criticality at AEET on July 10, 1960. CIRUS was well designed for producing weapons-grade plutonium.[21] The reactor is designed to achieve low-burn-up on the natural uranium fuel by quicker refueling sequences. This low burn-up on fuel produces plutonium of weapon-grade. The reactor is capable of producing approximately 9 kilograms of plutonium (as derived from the simulations carried out within the assumptions stated later in the book) in a year. The CIRUS reactor is not now nor has ever been under International Atomic Energy Agency (IAEA) safeguards.

[18] R. Wohlstetter, "The Buddha Smiles: Absent-Minded Peaceful Aid and the Indian Bomb,"Pan Heuristics, Los Angeles, p. 32 (1977); and G. Perkovich, "India's Nuclear Bomb: The Impact on Global Proliferation," University of California Press, Berkeley, p. 29, (1999).

[19] B, Chellaney, "Nuclear Proliferation: The US-India Conflict," Orient Longman, New Delhi, p. 36 (1993).

[20] The CIRUS reactor was based on the NRX reactor design in Canada and used a heavy water moderator and natural uranium fuel. Natural uranium consists of approximately 0.72 % U-235 and 99.28% U-238. The heavy water moderator is used to slow down the neutrons produced during fission to allow them to be absorbed in the small quantity of U-235 contained in the natural uranium fuel. Some portion of these neutrons are also absorbed in the U-238 and produce Pu-239, which is the primary isotope used in nuclear weapons. This reactor produced the plutonium used in India's first nuclear test in 1974. India argued in 1974 that the contract allows its use in producing peaceful nuclear explosives, which is how it characterized the 1974 explosion, though in recent years the project director Raja Ramanna has conceded that this was a sham. CIRUS also provided the design prototype for India's more powerful DHRUVA plutonium production reactor and is directly responsible for producing nearly half of the weapons-grade plutonium currently believed to be in India's stockpile.

[21] Weapons-grade plutonium is plutonium containing a low percentage of Pu-240 and a high percentage of Pu-239. Generally, speaking weapons-grade plutonium has a Pu-240/Pu-239 ratio of approximately 6%.

In July 1958, Prime Minister Nehru authorized project PHOENIX, a plan to build a spent fuel reprocessing facility[22] with a capacity of processing 20 tonnes of spent fuel a year. The capacity of PHOENIX was designed to match the production capacity of CIRUS.[23] The PHOENIX plant was based on the U.S. developed PUREX process and was commissioned in mid-1964. PHOENIX had an estimated capacity to separate up to 10 kg of plutonium annually.[24]

As a nation, India has always placed a premium on self-sufficiency. This traditionally closed nature of the Indian economy accounts for why the import/export based Chinese economy far outpaced the growth of the Indian economy from the late seventies to the early nineties. Due to its vast domestic resources of thorium but limited supplies of uranium, from the start of its nuclear program, India has always placed strong emphasis on the development of breeder reactor fuel cycles[25,26]. This provided a peaceful rationale for developing a plutonium separation capability, but the impetus for India's first spent fuel reprocessing plant had a lot to do with nuclear weapons option.

2.2   EARLY WEAPONS DEVELOPMENT EFFORTS: 1960-1966

The Indian political and scientific communities became aware of the Chinese nuclear program in the early 1960's. Several editorials in the Indian press in 1960 discussed the possibility of a possible Chinese nuclear weapons test.[27] Concerns over the Chinese weapons program spearheaded India's efforts toward weaponization beginning in 1961. At TIFR, Bhabha initiated preliminary studies on weapon physics using materials from the CIRUS reactor and the Trombay reprocessing plant. A physics group was established at TIFR in January 1962 to study the design of an implosion weapon. This group worked in secrecy and reported directly to Bhabha. The humiliating defeat by China in a border-dispute war in 1962 led to a formal demand for the development of nuclear weapons in the Indian Parliament.[28] Dr. Bhabha also argued in favor of matching the Chinese military power and

---

[22] Reprocessing facilities are used to chemically separate the plutonium from the uranium and fission products in spent fuel. Their capacity is generally listed in terms of the mass of spent fuel that can be processed annually in the plant. The quantity of plutonium then produced by the plant would depend upon the quantity of plutonium in each mass of spent fuel.

[23] http://nuclearweaponarchive.org/India/IndiaOrigin.html.

[24] Reprocessing facilities chemically separate the plutonium from the uranium and fission products in spent fuel. Their capacity is generally listed in terms of the mass of spent fuel that can be processed annually in the plant. The quantity of plutonium then produced would depend upon the quantity of plutonium in each mass of spent fuel.

[25] Breeder reactors are an advanced reactor design in which the reactor breeds additional fuel during operation. Technically, a breeder reactor produces more fuel during operation than it consumes. The breeder reactor achieves this through a design which highly conserves neutrons in the system. Typically, breeder reactors have been designed as fast reactor systems utilizing a driver fuel with a high fissile loading (such as high enriched uranium or plutonium) surrounded by a breeder blanket (consisting of depleted uranium or thorium).

[26] Thorium is a potential material for breeder reactors (breeding fissile U-233 from absorptions in Th-232). Breeder reactors require highly concentrated fissionable material for reactor fuel: either highly enriched uranium or plutonium and to extract this material from spent fuel it requires reprocessing capability.

[27] G.G. Mirchandani, India's Nuclear Dilemma, Popular Book Services, New Delhi, p. 13 (1968).

[28] S. Bhatia, India's Nuclear Bomb, Vikas Publishing, Ghaziabad, pp. 108-109 (1979).

even asked Prime Minister Nehru to authorize a nuclear test in Ladhak on the Chinese border.[29] This ambitious demand was, however, not supported by the existing Indian nuclear infrastructure. Not paralleling the urge for political support, the only plutonium production reactor (CIRUS) did not reach full power until October 16, 1963 and the PHOENIX plant at Trombay had a low utilization during the 1960's. PHOENIX was officially inaugurated on January 22, 1963 and began operation in April 1964, but it produced very little plutonium for the first several years. Thus, it is unlikely that India had the required plutonium for its first device until 1969.[30]

At the Pugwash Conference on Science and World Affairs held in Udaipur, India, Bhabha described how a state could use the aid provided by international organizations to bolster a nuclear program which could then be used for military purposes.[31] This is essentially the path followed by India in acquiring the technology for its nuclear weapons program. Following Nehru's death in May 1964 and the Chinese nuclear weapons test in October 1964, the strategy for the Indian nuclear weapons program was outlined by Bhabha and others. This strategy was openly debated in the Indian newspapers and on Indian radio.[32] In November 1964, the Indian policy shifted and focused on the development of a "peaceful nuclear explosive" (or PNE). Bhabha had previously acknowledged that there is no difference between a PNE and a nuclear weapon. The nuclear weapons program continued under the excuse of a PNE through the first weapons test in 1974 and all the way until the 1998 tests (when India finally acknowledged the actual objective of the program).[33]

2.3   NEGOTIATIONS AND AFTERMATH OF THE NON-PROLIFERATION TREATY: 1966 – 1974

India began as a strong advocate of the Non-Proliferation Treaty (NPT) but a turn of events made it a non-signatory of the final treaty. India insisted on universal disarmament with the non-existence of permanent nuclear powers. This fundamental stand lead India to vote against it on June 12, 1968.

During December 1968 to January 1969, a team of Indian scientists visited the Soviet Union. They were impressed by the plutonium-fueled, pulsed fast reactor for developing an excellent laboratory model for study of fission bomb. This was the type of reactor being used during the Manhattan Project (Dragon experiments) and later (Godiva, Popsy and Topsy reactors). Indian scientists subsequently designed PURNIMA (Plutonium Reactor for Neutron Investigation in Multiplying Assemblies), which was comprised of a hexagonal core of 177 stainless steel pencil shaped rods containing 18 kg of plutonium as 21.6 kg of plutonium oxide pellets. PURNIMA went critical on 18th May 1972 after sufficient separated plutonium finally became available. PURNIMA provided

[29] G. Perkovich, India's Nuclear Bomb: The Impact on Global Proliferation, University of California Press, Berkeley, pp. 60-62 (1999).
[30] http://nuclearweaponarchive.org/India/IndiaOrigin.html.
[31] H.J. Bhabba, "The Implications of a Wider Dispersal of Military Power for World Security and the Problem of Safeguards," Proceedings of the 12th Pugwash Conference on Science and World Affairs, Udaipur, India, pp. 75-80 (1964).
[32] Numerous references to this debate could be given from the Statesman, the National Herald, and the Indian Express; however, perhaps of more interest to the reader is the detailed discussion of this decision given by G. Perkovich on pages 64–85 in India's Nuclear Bomb: The Impact on Global Proliferation.
[33] http://nuclearweaponarchive.org/India/IndiaOrigin.html.

the test bed for understanding the physics of fast fission and neutrons. On the footings of PURN-IMA, the scientists developed facilities and gained experience in handling of plutonium. Work also began on fabricating plutonium metal alloys for construction of a bomb core. To develop the essential implosion system, scientists performed numerical implosion simulations on Soviet Besm 6 computer.[34]

The basic design for India's first nuclear device was complete by early 1972. The important decision of assembling the device and preparing for the test was made later in 1972. The decision was heavily influenced by internal momentum and domestic politics. In order to not alert proliferation observers, Homi Sethna, head of BARC, separated the Indian space program from the DAE. This eliminated the DAE from developing both nuclear explosives and missile technology. The arguments for the bomb were stated as "without it India could not expect to be admitted to the corridors of global power, nor enjoy the status of the dominant regional power; that the bomb might quicken the process of normalizing relations with China; that it would proclaim India's independence of the Soviet Union and compel the United States to change its attitude of hostility or benign neglect."[35]

In 1970, the PHOENIX plant developed a serious leak and had to be shut down for maintenance. This limited the material available for a weapon. To provide the material for a test device, the PURNIMA reactor was shutdown in January 1973 and dismantled. PURNIMA contained 18 kg of plutonium and it is assumed that the test device required about 6 kilograms of plutonium. Therefore, in 1974 there was only enough material for three devices. The test was named the "Smiling Buddha" and was successfully conducted on May 18, 1974, at Pokhran, in Rajastan. The material used in the device was produced originally by CIRUS but was first used in PURNIMA. In the aftermath of the test, the civilian nuclear power program struggled for the next three decades due to lack of domestic resources and unavailability of international technology because of the nuclear isolation stemming from the test.[36]

2.4   INDIA'S ISOLATION FOLLOWING OPERATION SMILING BUDDHA: 1975–1998

International reaction to the "Smiling Buddha" was varied. The 1974 test sharply escalated international attention to nuclear proliferation, and international support for India's civilian nuclear program disappeared. Canada cut off virtually all nuclear assistance four days after the test, bringing two nuclear power projects (the RAPS-II reactor and the Kota heavy water plant) to a halt. In 1978, but as a direct consequence of the 1974 test, the U.S. congress passed amendments to the U.S. Atomic Energy Act of 1954 that served the purpose of prohibiting U.S. involvement with India's nuclear sector, albeit without specifically naming India. India had sufficient natural uranium to complete the RAPS-II reactor and met startup and refueling needs; however, all future PHWR projects were seriously hindered.

---

[34] http://nuclearweaponarchive.org/India/IndiaOrigin.html.
[35] Bhabani Sen Gupta, "Nuclear weapons? Policy options for India" Sage, New Delhi (1983).
[36] http://nuclearweaponarchive.org/India/IndiaOrigin.html

In 1977, India started work on a larger 100 MW_{th} plutonium production reactor at Trombay. This reactor was named "DHRUVA." There were concerns over heavy water supplies for DHRUVA and these concerns also existed for the civilian reactors. MAPS-I also suffered due to a lack of heavy water availability. On August 8, 1985, DHRUVA went critical but was soon crippled by serious problems requiring a shutdown. It was restarted in December 1986 and operated at 25% capacity factor from then until spring of 1987. DHRUVA finally achieved full power on January 17, 1988. After that time, DHRUVA became the main supplier of plutonium for the Indian nuclear weapons.[37]

The arguments for developing the weapons can be assessed from a statement by then Prime Minister I.K. Gujral in a meeting with President Bill Clinton at the U.N. General Assembly session in September 1997, "I told President Clinton that when my third eye looks at the door of the Security Council chamber it sees a little sign that says 'only those with economic power or nuclear weapons are allowed.' I said to him, 'it is very difficult to achieve economic wealth'."

2.5   POKHRAN-II TESTS IN 1998

India conducted a second series of nuclear weapon tests in May 1998. These tests were also conducted at the Pokhran facility and were referred to as the Pokhran-II tests. The Pokhran-II tests were conducted on the same day (May 11) on which the 1974 test was conducted. There are conflicting reports on the claimed yields and design of these devices. Some reports claim that one of the devices used reactor-grade plutonium, which created a furor as this is presently and abundantly produced by CANDU reactors.

2.6   FOLLOWING THE POKHRAN II TESTS: 1999 - 2006

By 1998, India had 8 PHWR's of 220 megawatt electrical (MWe) power rating. The share of electricity generation that came from nuclear power had increased gradually but was well below expectations. It was nowhere close to the planned generation of 20,000 MWe by 2000. By March of 2006, India had an installed capacity of 3900 MWe from nuclear power plants. Nuclear energy comprised of 3% of the total installed capacity.[38]

The plutonium from the spent fuel of production reactors was reprocessed for use in the Fast Breeder Test Reactor (FBTR) of 40 MW_{th}, which entered service in 1985. This technological demonstration opened up the doors for establishment of two Prototype Fast Breeder Reactors (PFBR) of 500 MWe each, which is presently planned to enter operation in 2010 and 2012.[39]

A research reactor of 30 kW_{th} rating (KAMINI) having U-233 as driver fuel and fixed along with adjustable BeO reflectors was built and operated to demonstrate the feasibility of the

---

[37] http://www.barc.ernet.in/webpages/reactors/dhruva.html, "DHRUVA Reactor," Bhabha Atomic Research Center, Mumbai (August 2007).

[38] http://www.npcil.nic.in/profile_eng25jan06.pdf, "Profile," Nuclear Power Corporation of India Limited, India (2006).

[39] D. Ramana et.al, "The Indian Nuclear Tests – Summary Paper," Bharat Rakshak Monitor, Vol. 3(6) (May-June 2001).

third-stage.[40] Furthermore, with this reactor, a foundation to conduct research for development of commercial-scale U-233 fueled reactors was established.

Immediately after the Pokhran-II tests, India faced the dual challenges of international sanctions and diminishing uranium reserves from the flagship mine in Jaduguda. Prior to that date, the constraints on uranium fuel production were milling capabilities. The affect of international sanctions was evident in the power rating and number of reactors. India had planned to generate 20,000 MWe by year 2000 but could only realize 15% of the proposed. the power ratings of all the power reactors never exceeded 220 MWe though two units of 540 MWe PHWR's were in advanced commissioning stage. After the Pokhran-II tests, the focal point shifted to the domestic uranium ore reserves. It has been observed through multiple open sources that India struggled to meet the fueling needs of the existing reactors and had to decrease the electrical output to match the fuel availability. There were cases of social unrest coupled with diminished ore sources in the Jaduguda mine leading to a drop in ore production. The newly found mines and formerly abandoned mines failed to match the demand because of lower ore grade, delay in the part of local authorities and social unacceptability. Although mining activities at these sites were rigorously pursued, some were abandoned due to the lower ore yield resulting in economic burden from these sites, along with political and social reasons.[41]

Following Pokhran-II tests, the consequences of sanctions may not have directly altered the strategic and civilian nuclear programs, but they contributed to the compelling possibility of a scenario in which production of uranium ore dropped substantially, thereby stalling further nuclear power generation program.

2.7   CURRENT STATE OF THE INDIAN NUCLEAR PROGRAM

There are no official figures available of weapon stockpiles at any developmental stage of India's arsenal. Conclusions can only be inferred from considerations of India's ability to produce critical raw materials and production plants.

India's first-stage nuclear power program of CANDU reactors has, since 1978, proceeded largely without fuel or technological assistance from other countries. Partly as a result of this, its power reactors have been among the worst-performing in the world (at least with regard to capacity factors). This reflects the difficulties of technological isolation. However, the capacity factors apparently improved in late 90s and newer projects were constructed at a faster rate. This pace, though, could not live longer because of the diminishing uranium reserves. India, though, continued the research and development of higher power rated CANDU reactors along with second stage fast breeder reactor designs. Governmental support never ceased for the constructions of the planned PHWR's (220 MWe and 540 MWe) and PFBR (500 MWe) at Kalpakkam (in the southern state of

---

[40] C. S. Sunny and et.al., "KAMINI reactor benchmark analysis," annals of Nuclear Energy, India (2007).

[41] A. B. Awati and R. B. Grover, "Demand and Availability of Uranium resources in India," Department of Atomic Energy, Mumbai, India (2005).

Tamilnadu). Extensive negotiations were pursued with Russia for at least two units of 1000 MWe reactors.

This new phase of events followed its course with the proposal for ending civilian nuclear technological isolation by allowing international safeguards on some of the existing PHWR's along with all the future PHWR's. The next three chapters consolidate all the segments of the nuclear fuel cycle and build a roadmap of the possibilities India is facing, following this nuclear accord.

CHAPTER 3

Indian Nuclear Facilities

India has a large and diverse nuclear power program which essentially includes all components of a closed nuclear fuel cycle. In this chapter, a description of the existing and planned Indian nuclear facilities is given. These facilities stand as an epitome of possibilities that were realized starting with international technological support and limited domestic resources. India had been extremely forthcoming with details of almost all the aspects of the nuclear program with the exception of uranium production. Although the capacity of milling and conversion process along with fuel fabrication facility was available through open source information. Starting with the briefing of the nuclear power plants, a CANDU fuel bundle was modeled to analyze the fuel configuration. The input code of the fuel design was simulated for exact operating conditions and the prerequisite information was outputted through an output code. The information of the fuel isotopic for different groups of neutron energies and criticality at multiple burn-up steps was thoroughly analyzed in Chapter 4 to interpret the uranium usage. Declared capacity of the nuclear power plants and a pragmatic assumption of the plutonium production reactors operability along with probable capacity of the uranium enrichment plant was the basis for computing the usage of natural uranium at three specific timelines. At every step on the reverse time frame the feasibility of uranium production given the capacity of milling, presence of ore and the handling capacity fuel fabrication facility was correlated to develop the block diagram flowsheet of the nuclear fuel cycle. The details and results of the study have been stressed upon in Chapters 4 and 5 following the assimilation of the Indian nuclear facilities in this chapter.

3.1 POWER REACTORS

The first nuclear power project of India started with General Electric constructing and commissioning two Boiling Water Reactor (BWR) power plants at Tarapur (near Mumbai, formerly Bombay) in 1969. Soon India realized the difficulty in acquiring enriched uranium for these reactor types and believed that BWR's would ensure lifetime dependence on the U.S. for fuel needs. India received the power plants at a meager cost of 80 million U.S. dollars. At the time, the GE-built BWR's were the biggest electricity producing units in India and played a primary role in the development of Bombay, the economic capital of India. Even before India's first nuclear power plant at Tarapur, Homi Bhabha and his team were suggesting the three-stage-program discussed in Chapter 2 and looked into the potential of CANDU-type reactors. This carried the burden of acquiring heavy water for moderation and as a coolant but using indigenous natural uranium fuel. Apparently the technology for production of this fuel existed in India.

Prior to 1998, there existed 12 nuclear power plants in India with 8 of them outside safeguards. Among the plants under safeguards, two were RAPS-1 & 2 and the Tarapur BWR's, which were under safeguards in reciprocity for fuel provision from a sequence of international fuel sources. The technology to build 220 MWe rated units was mastered and plans were laid to further the progress of nuclear power plants. Building of nuclear power reactors continued after the Pokhran-II tests with the addition of six more plants including two of 540 MWe ratings. However, until the 540 MWe Tarapur-4 unit in 2005, India did not have a nuclear power producing unit greater than 220 MWe. Old reprocessing plants are being scaled up and new ones were built to meet the fuel needs of the breeder reactors.

India currently (late 2008) possesses 15 operational PHWRs and 2 operational BWRs. 3 PHWRs of 220 MWe along with 2 VVERs of 1000 MWe are under construction. India's operational nuclear power plants have 4120 MWe of generating capacity. There, currently, is a strong initiative to construct two power reactors per year. Table 3.1 lists all the existing Indian nuclear power plants, their power rating, type and date of commencement into service.[1]

India, being alienated from the advanced nuclear technology that it badly needed, was faced with technological hurdles following the 1974 weapon test. Over a period of time, India obtained self sufficiency in its indigenous PHWR nuclear power plant technology, but until recently, all the nuclear power plants were rated at 220 MWe and operated at a low capacity factor. Capacity factors of the order of 80% were achieved in 2003 from newly commissioned power plants. Table 3.2 lists the capacity factors for the existing nuclear power plants until 2003. Table 3.3 extends the capacity factor data in a consolidated form to 2006. These capacity factors were estimated as follows: The aggregate energy (MWe) generated by each of the plants was estimated for the requisite periods, based on NPCIL data.[2] Start dates for the various plants were obtained similarly, and capacity factor was accordingly calculated on the basis of the rated powers shown in Table 3.1. These capacity factors agree well, but not exactly, with those reported by the NPCIL (*ibid.*).

India has an aggressive fast breeder reactor program; however, India does not currently have an operating fast breeder power reactor. Technology demonstration was performed with the Fast Breeder Test Reactor (FBTR). Given the experience from the FBTR, construction was started on 500 MWe Prototype Fast Breeder Reactors (PFBR). The reactor is expected to be operational in 2010. Indira Gandhi Center for Atomic Research (IGCAR) has expressed its desire to construct 4 more PFBRs by 2020.[3]

---

[1] Data for the Indian power reactors were taken from several sources, including D. Albright and S. Basu, "Separating Indian Military and Civilian Nuclear Facilities," Institute of Science and International Security (2005). Available at: http://www.isis-online.org/publications/southasia/indiannuclearfacilities.pdf; The IAEA Power Reactor Information System database (http://www.iaea.org/programmes/a2/index.html); The official website of the Nuclear Power Corporation of India Limited http://www.npcil.nic.in/main/aboutus.aspx; and from J. Cirincione, Deadly Arsenals: Tracking Weapons of Mass Destruction, Carnegie Endowment for International Peace, Washington, D.C. (2002).

[2] NPCIL, "Plants Under Operation," http://www.npcil.nic.in/main/PowerPlantDisplay.aspx, accessed April 30, 2009.

[3] The Hindu, September 07, 2005, http://www.hinduonnet.com/2005/09/07/stories/2005090704781300.htm.

Table 3.1: List of nuclear power plants of India (*Continues*).			
Plant Name	**Type**	**Rated Power (MWe)**	**Commercial Operation Started**
Kaiga Atomic Power Station-1 (KAIGA-1)	PHWR	220	16 November, 2000
Kaiga Atomic Power Station-2 (KAIGA-2)	PHWR	220	16 March, 2000
Kaiga Atomic Power Station-3 (KAIGA–3)	PHWR	220	6 May, 2007
Kaiga Atomic Power Station-4 (KAIGA-4)	PHWR	220	March, 2009
Kaiga Atomic Power Station-5 (KAIGA–5)	PHWR	220	Proposed for construction
Kaiga Atomic Power Station-6 (KAIGA–6)	PHWR	220	Proposed for construction
Kakrapar Atomic Power Station-1 (KAPS-1)	PHWR	220	6 May 1993
Kakrapar Atomic Power Station-2 (KAPS-2)	PHWR	220	1 September, 1995
Kundankulam Power Plant-1 (KK–1)	VVER	1000	Planned for August, 2009
Kundankulam Power Plant-2 (KK-2)	VVER	1000	Planned for May, 2010
Madras Atomic Power Station-1 (MAPS-1)	PHWR	220	27 January, 1984
Madras Atomic Power Station-2 (MAPS-2)	PHWR	220	21 March, 1986
Narora Atomic Power Station-1 (NAPS-1)	PHWR	220	1 January, 1991
Narora Atomic Power Station-2 (NAPS-2)	PHWR	220	1 July, 1992
Prototype Fast Breeder Reactor-1 (PFBR-1)	FBR	500	Planned for 2010
Prototype Fast Breeder Reactor (PFBR)	FBR	500	Proposed for construction by 2020
Prototype Fast Breeder Reactor (PFBR)	FBR	500	Proposed for construction by 2020

Table 3.1: (*Continued*) List of nuclear power plants of India.

Plant Name	Type	Rated Power (MWe)	Commercial Operation Started
Prototype Fast Breeder Reactor (PFBR)	FBR	500	Proposed for construction by 2020
Prototype Fast Breeder Reactor (PFBR)	FBR	500	Proposed for construction by 2020
Rajasthan Atomic Power Station-1 (RAPS-1)	PHWR	100	16 December 1973
Rajasthan Atomic Power Station-2 (RAPS-2)	PHWR	200	1 April 1981
Rajasthan Atomic Power Station-3 (RAPS–3)	PHWR	220	1 June 2000
Rajasthan Atomic Power Station-4 (RAPS-4)	PHWR	220	23 December 2000
Rajasthan Atomic Power Station-5 (RAPS–5)	PHWR	220	Planned for February 2009
Rajasthan Atomic Power Station-6 (RAPS-6)	PHWR	220	Planned for June 2009
Rajasthan Atomic Power Station-7 (RAPS–7)	PHWR	540	Proposed for construction
Rajasthan Atomic Power Station-8 (RAPS–8)	PHWR	540	Proposed for construction
Tarapur Atomic Power Station-1 (TAPS-1)	BWR	160	28 October 1969
Tarapur Atomic Power Station-2 (TAPS-2)	BWR	160	28 October 1969
Tarapur Atomic Power Station-3 (TAPS–3)	PHWR	540	18 August 2006
Tarapur Atomic Power Station-4 (TAPS-4)	PHWR	540	12 September 2005

Table 3.2: Capacity factors of PHWRs until 2003.

Plant Name	Capacity Factor (%)
RAPS-1	23.31
RAPS-2	52.65
MAPS-1	52.82
MAPS-2	52.92
NAPS-1	60.62
NAPS-2	67.82
KAPS-1	70.91
KAPS-2	84.14
KAIGA-1	80.70
KAIGA-2	80.91
RAPS-3	77.98
RAPS-4	79.20

Table 3.3: Annual average capacity factors of PHWRs from 2004 to 2006.

Power Plant	Capacity Factor/Year
All 12 Plants listed in Table 3.2	81%/2004
TAPP-4 + 12 Plants	76%/2005
TAPP-3, TAPP-4 + 12 Plants	52.4%/2006

3.2 RESEARCH AND PLUTONIUM PRODUCTION REACTORS

India has a number of research and plutonium production reactors. Construction of research reactor facilities began with the APSARA and CIRUS reactors, which were already mentioned in the preceding chapter. India's research reactor program has continued vigorously to this day. To advance research on the development of the second-stage power reactor systems, the FBTR was built based on the French Rapsodie design. This 40 MW_{th} fast reactor was made operational with a mix of plutonium and uranium carbide as fuel.

Technological demonstration of U-233 based reactor, which is crucial to the third stage of the three-stage program, was implemented with the commissioning and operation of the 30 kW_{th} KAMINI reactor. However, commercial systems have yet to be demonstrated because of non-existence of U-233 fuel in mass scale and viable power reactor design.

India constructed an additional weapons-grade-plutonium-producing reactor and numerous other research reactors for neutronics studies. A nuclear reactor designed to achieve high conversion ratio along with rapid on-power refueling sequences or quicker off-power fuel changes would result in low burn-up on fuel. Given the plutonium isotopes ratio of less than 6% for Pu-240 to Pu-239 for weapons-grade plutonium classification, the low burn-up fuel meets the strict criteria. Table 3.4 lists all the research reactors, including their capacity, type, date of commencement into service and their function.[4]

Table 3.4: Indian plutonium production and research reactors.				
Name	Location	Type	Start Date	Function
CIRUS	Trombay	40 MWth HWR	10 July, 1960	Plutonium Production
DHRUVA	Trombay	100 MWth HWR	10 August, 1985	Plutonium Production
APSARA	Trombay	1 MWth LWR	1956	Research and Development
PURNIMA-1	Trombay	Critical Assembly	1989	Decommissioned
PURNIMA-2	Trombay	LWR	1984	Decommissioned
PURNIMA-3	Trombay	LWR	1994	Uses U-233
Zerlina	Trombay	PHWR	1961	Decommissioned
KAMINI	Kalpakkam	30 KWth	1996	Uses U-233
Andhra University	Vishakapatnam	0.1 MWth	Unknown	Research and Development
FBTR	Kalpakkam	40 MWth FBR	1998	Fast Breeder Development

3.3   URANIUM ENRICHMENT FACILITIES

India's large scale uranium enrichment endeavor started in the early 1990's. The purpose of uranium enrichment in India is speculated to be for military (naval) applications and to provide fuel for the fast breeder reactors. Table 3.5 lists the enrichment facilities, including their location, types of technology and date of commencement into service.[5] Of these facilities, only the Mysore facility is capable of large-scale operation. In estimating the enriched uranium accumulation for the Mysore facility, P1 centrifuges of 3 SWU/yr capacities with a total plant load of 2000 SWU/yr were assumed.

---

[4] Data for the Indian research reactors was taken from several sources including: the IAEA Research Reactor Database (http://www.iaea.org/worldatom/rrdb/); the Bhabba Atomic Research Center Webpage (http://www.barc.ernet.in/); D. Albright and S. Basu, "Separating Indian Military and Civilian Nuclear Facilities," Institute of Science and International Security (2005); and from J. Cirincione, Deadly Arsenals: Tracking Weapons of Mass Destruction, Carnegie Endowment for International Peace, Washington, D.C. (2002).

[5] D. Albright and S. Basu, "Separating Indian Military and Civilian Nuclear Facilities," Institute of Science and International Security, Washington D.C. (Dec 19, 2005).

There were reports claiming failure of this project because of its inability to produce weapons-grade enriched uranium.[6]

Name	Location	Type	Start Date	Function
Center for Advanced Technology	Indore	Laser Enrichment	1993	Research
Rare Materials Project	Mysore	Centrifugal	1991	Uranium Enrichment
Laser Enrichment Plant	Trombay	Laser Enrichment	1993	Research
Uranium Enrichment Plant	Trombay	Pilot Scale Ultracentrifuge	1985	Research & Development

Table 3.5: Indian enrichment facilities.

3.4   HEAVY WATER PRODUCTION FACILITIES

Table 3.6 lists the heavy water production units with their location and start date for each facility. The heavy-water production technology of India is outdated and large-scale renovation work, with international collaborations, is needed. Heavy-water production in India supports domestic PHWR nuclear power plants. Delays in nuclear power plant construction in the past and low capacity factor operation at present had not strained the heavy water production, but with increasing demand, the scenario may be different. There are, in total, eight heavy-water production plants with a capacity of more than 650 tonnes/yr, adequate both to support current domestic requirements and export sales. Six of these plants operate on ammonia exchange processes and two on the hydrogen sulfide process. [7]

3.5   FUEL FABRICATION FACILITIES

All fuel fabrication facilities are listed in Table 3.7.[8] The Nuclear Fuel Complex (NFC) at Hyderabad is the only large scale CANDU fuel fabrication facility in India. The NFC has an annual handling capacity of 250 tonnes of UF_6 and the estimated capacity as input to the plant is 216 tonnes of UF_6, after losses. In 2006, the NFC raised its handling capacity from 250 to 600 tonnes of UF_6 per year.[9]

---

[6] M.V. Ramana, "India's Uranium Enrichment Program," INESAP Information Bulletin 24 (Dec 2004).

[7] http://www.globalsecurity.org/wmd/world/india/hazira-nuke.htm

[8] D. Albright and S. Basu, "Separating Indian military and Civilian Nuclear Facilities," Institute of Science and International Security, Washington D.C. (Dec 19, 2005).

[9] Data for the installed and increased capacity of the Nuclear Fuel Complex was taken from the website of the Nuclear Threat Initiative, "India Profile, Nuclear Facilities, Nuclear Fuel Complex" Washington D.C. (September 2003) (http://www.nti.org/e_research/profiles/India/Nuclear/2103_2487.html). An estimation of tonnes of UF_6 inputted was computed after considering the isotopic conversion along with losses inquired in the process.

Table 3.6: Indian heavy water production plants.

Name/Location	Process Type	Commissioned On
Baroda	Ammonia Exchange	1980
Hazira	Ammonia Exchange	1991
Nangal	Ammonia Exchange	1962
Talchar	Ammonia Exchange	1985
Thal - Vaishet	Ammonia Exchange	1985
Tuticorin	Ammonia Exchange	1978
Kota	Hydrogen Sulfide	1981
Manuguru	Hydrogen Sulfide	1991
Total capacity		650 tonnes per year

The higher capacity though can only cater to the need of 14 PHWRs operating at 90% capacity factor. Any further addition of PHWR's would require additional rise in capacity. Presently, the NFC production is not a restriction to the nuclear power plant operation because of the lower operational capacity factor due to reduced uranium reserves. The scenario may change with additional mines or international uranium (in the form of UF_6) derived from the cooperation agreement.

3.6   REPROCESSING FACILITIES

Table 3.8 lists all of the reprocessing facilities in India.[10] These facilities are necessary for both weapons-grade plutonium production and for the second stage of the Bhabha three-stage program. Kalpakkam's 100 tHM/yr reprocessing facility is the latest in bridging the gap between India's (Bhabha) first-stage and second-stage power program. This facility would meet the fuel needs of the under-constructed PFBR and the proposed fast breeders.

3.7   URANIUM MINING, MILLING, AND CONVERSION FACILITIES

Production of uranium typically begins with mining of some uranium-bearing ore. For purposes of a quantitative estimate of uranium production from a mine three parameters are of key importance:

First is the "production capacity" of the mine, by which is intended the rate of production of material from the mine, under ideal conditions. Here production capacities typically will be denominated in metric tonnes of material per (working) day.

Second is the "grade" of the ore. This is the percentage, by weight, of the ore that is uranium. Grades may be denominated in terms of weight percentage of the element uranium, or of one of its stable compounds, most often U_3O_8, which often is termed "yellowcake."

---

[10] D. Albright and S. Basu, "Separating Indian military and Civilian Nuclear Facilities," Institute of Science and International Security, Washington D.C. (Dec 19, 2005).

Table 3.7: Indian fuel fabrication facilities.

Name	Location	Type	Capacity (tHM/yr)	Start Date	Function
Enriched Fuel Fabrication Plant	Hyderabad	BWR	25	1974	LWR Fuel Assemblies (Safeguarded)
Advanced Fuel Fabrication Facility (AFFF)	Tarapur	Unknown	20	1990	MOX Fuel for BWR, PFBR, PHWR & Research & Development
Nuclear Fuel Complex (NFC)	Hyderabad	PHWR	250	1971	PHWR Fuel Bundles
			600	2006	
MOX Breeder Fuel Fabrication	Kalpakkam	Pilot Scale	Unknown	Unknown	MOX Fuel

The third key mining parameter is the "average mining recovery," which means the percentage of the material extracted from the mine that is actually (uranium-bearing) ore.

The next step in production is the extraction of U_3O_8 from the geological ore. This involves massive mechanical and chemical processes, which are carried out at a milling facility. The ore is first crushed, then roasted and quenched to remove organics, followed by leaching to scrap out the solid waste. Approximately 15-20% loss of U_3O_8 typically is incurred in these processes; the fraction retained, termed as "average process recovery" is one of the important parameters characterizing a mill. The other is mill size (capacity), which here will be denominated in the same terms as production capacity, tonnes (of material) per day (that could be processed if sufficient supply of material were available).

Finally, if the uranium is to be enriched, then the U_3O_8 typically is later converted to the stable gaseous compound UF_6. Parameters relevant to characterizing a conversion facility are discussed further in the following chapter, as estimates of the amount of enriched uranium produced by India are developed.

Exploration for uranium ore in India started practically coincident with independence in 1947. Significant production seems to have begun around 1967. At that time the Nuclear Non-Proliferation Treaty (NPT) was being formulated and steps were taken for nonproliferation of nuclear materials

Table 3.8: Indian reprocessing facilities.

Name	Location	Type	Capacity (tHM/yr)	Start Date	Function
Power Reactor Fuel Reprocessing Plant (PREFRE)	Tarapur	PUREX	100 150	1977 1991	Reprocess CIRUS, DHRUVA & PHWR fuel. Provide fuel for FBTR & AFFF
Kalpakkam Reprocessing Plant (KARP)	Kalpakkam	PUREX	100	1998	Reprocess MAPS & FBTR fuel. Provide fuel for PFBR
Fast Reactor Fuel Reprocessing Plant (FRFRP)	Kalpakkam	Full Scale	Unknown	Future	Reprocess FBTR fuel. May provide fuel for PFBR
Lead Minicell Facility	Kalpakkam	Demonstration	Unknown	2003	Reprocess FBTR & PFBR fuel in future
Plutonium Reprocessing Plant	Trombay	PUREX	30 50	1964 1984	Reprocess CIRUS, DHRUVA fuel for weapons-grade plutonium

and technology. Uranium mining was initiated with governmental backing. Beginning with Jaduguda (located in the eastern part of India), six to seven different mining locations were discovered over a period of time. Table 3.9 contains key parameters for all sites that seem to have been exploited through 2006.

Except as explicitly otherwise, noted either in the associated textual discussion or a footnote, these data are taken from an appropriate edition of the IAEA Red Book.[11] When parameters vary between editions, the largest encountered value is listed. Unfortunately, the Red Book began providing estimates of production at individual mines only with its 1999 edition. Therefore, in some cases it is necessary to resort to other means of estimating the historic maximum production, which is used here as production capacity. This will be discussed further, in connection with the subsequent discussion of the individual mines. Unfortunately, the Red Book does not cite ore grades, so those values necessarily are taken from elsewhere, as indicated by individual citations. When judgments are necessary, in order to reconcile discrepant values, overestimation of India's capacity to produce uranium has been made.

The Jaduguda mine, located in the Singhbum East District of Jharkand (formerly the southern portion of undivided Bihar state), began operations in 1967. It was long the primary source of Indian uranium. The estimate of production capacity of this mine that is given in Table 3.9 is actually the size (capacity) of the mill also located at Jaduguda, which carries out milling activities for the Jaduguda mine, and also the nearby Bhatin and Narwapahar mines. (See below for more details of this milling facility.) The maximum production indicated in any *Red Book* is 1000 metric tonnes per day (in the 1973 edition).

---

[11] The most recent instance is Uranium 2007: Resources, Production and Demand, NEA No. 6098, OECD Nuclear Energy Agency and the International Atomic Energy Agency, Paris and Vienna. Hereafter this will be referred to as "Red Book: 2007," and similarly for other editions.

Table 3.9: Key parameters for various Indian uranium mining locations.

Location	Time Frame	Production Capacity (metric tonnes/day)	Estimated Ore Grade (%U)	Average Mining Recovery (%)
Jaduguda	1967–2006	1000	0.06[a]	80
Bhatin	1986–2006	250	0.06[b]	75
Narwapahar	1995–2006	1000	0.05[c]	80
Turamdih	2003–2006	550	.025[d]	75

[a] This estimate is a rough mean of values cited in a number of sources. For some examples, S. Jha, A.H. Khana and U.C. Mishrab, "A study of the technologically modified sources of 222Rn and its environmental impact in an Indian U mineralised belt," Journal of Environmental Radioactivity 53 (2001) 183-197, estimate 0.05-0.06% U_3O_8. This equates to .042-051% U. The WISE Uranium Project web site (http://www.wise-uranium.org/uoasi.html#IN, accessed March 7, 2009) cites 0.04% U. A.K. Singh et al., Applied Radiation and Isotopes 51 (1999) 107-113 say, of the Jaduguda mine, "The grade of the ore increases from an average of 0.067% U_3O_8 to about 0.20% (Bhola, 1971)." The quoted average value equates to 0.057% U.

[b] T.S. Subramanian, "Uranium Production From ore to yellow cake" Frontline, September 10, 1999 (http://www.dae.gov.in/press/spg/ucilfl.htm, accessed March 7, 2009) says "Bhatin's uranium deposit is the western extension of the Jaduguda ore body." On this basis the ore grade for Bhatin is here taken to be the same as that for Jaduguda.

[c] C. Ghosh, J.B. Narasimhan and K.K. Majumdar, "Studies in the Beneficity of Uranium Minerals from Narwapahar-Singhum", Received on April 29 (1969), Bhabha Atomic Research Center, Bombay, (http://www.new.dli.ernet.in/rawdataupload/upload/insa/INSA_1/20005acd_285.pdf, accessed March 8, 2009) cite 0.055% U_3O_8 = .047% U. The WISE Uranium Project web site (http://www.wise-uranium.org/uoasi.html#IN, accessed March 8, 2009) cites 0.05% U. Singh et al. (op cit. ola, 1971) quote (Bhola, 1971) to the effect of a grade of .058% U_3O_8 (.049% U) at Narwapahar.

[d] A. Mishra et al., "Microbial recovery of uranium using native fungal strains," Hydrometallurgy 95 (2009), 175–177, indicate 0.03% U_3O_8 = .025% U.

The Bhatin mine is much smaller than the nearby Jaduguda mine. It shares the majority of its infrastructure with Jaduguda. The Bhatin mine has been in operation since 1986.

The Narwapahar uranium mine and mill became operational in 1995. There are sources that indicate this mine was constructed with the assistance of Russian technology, but the genesis of those reports is unclear. The 1997 Red Book described this mine as "under construction;" however, from the1999 edition and later the Red Books have consistently listed a start date of 1995. These later Red Books have also consistently estimated the size (capacity) of the Narwapahar mine as 1000 tonnes of ore per day. An article dated September 10, 1999 states "the Narwapahar mine …now produces about 650 tonnes of ore a day; it is planned to increase production to about 1000 tonnes."[12]

The Turamdih mine was first mentioned in the 2001 Red Book, as "under development." In the 2005 edition this mine is indicated as having been "commissioned" in 2003, and that is also listed as the "start-up date." The Narwapahar and Turamdih mines have been described[13] as "the most modern mines in the country, highly mechanized with trackless vehicles used for movement of men and material."

India historically has produced some uranium by processing tailings from copper mines. However, the maximum production rate seems unlikely to have ever exceeded 5 metric tonnes per year. This contribution is, therefore, neglected here, as its maximum possible magnitude seems well within the errors in the present estimates of production from uranium mines. Uranium can be extracted as a by-product of other processes for recovering mineral resources, and there has been[14] speculation that "India may be obtaining nearly half of its uranium from secondary sources."

Prospective mining projects cited in the 2007 Red Book include Bagjata (start-up date 2007, size 500 tonnes ore per day, 80% average mining recovery), Banduhurang (2007, 2400, 65%), Mo-huldih (2011, 500, 80%), Lambapur-Peddagattu (2012, 1250, 75%), Tummalpalle (2010, 3000, 60%) and Kylleng-Pyndengsohiong-Mawthabah (2012, 2000, 90%).

As already indicated, mining is followed by milling for recovering U_3O_8 from the ore. As of 2006, there were two existing processing plants for milling.

The Jaduguda processing plant serves the Jaduguda, Bhatin and Narwapahar mines. ("The ore excavated from mines is transported by road to the ore processing plant, i.e., the uranium mill at Jaduguda."[15]) It was commissioned in 1967, with a capacity of 1370 tonnes of ore per day (*Red Book: 1997*). It has been stated that around 1998 it expanded to its present capacity of 2100 tonnes per working day;[16] however, in the following chapter some doubts are raised regarding such assertion. The maximum process recovery cited for this facility is 95%, in the 1999 Red Book. That figure has

---

[12] T. S. Subramanian (2005), op cit.

[13] Sutapa Bhattacharya, "Safety and Environmental Aspects in Processing of Uranium," AERB Newsletter, Vol. 18, No. 2-3, April-September 2005.

[14] M. M. Curtis, "India's Worsening Uranium Shortage," Report No. PNNL-16348, Pacific Northwest National Laboratory, January 2007.

[15] Sutapa Bhattacharya (2005), op cit.

[16] The 1997 Red Book mentions "an installed capacity of 1370 t/day ore," but indicates further that "the mill is being expanded to process 2000 t/day ore." The 1999 edition cites a "size" of 2100 tonnes ore per day.

subsequently, consistently declined, to 80% in the 2005 edition. The Jaduguda processing plant also will serve the Bagjata mine, when the latter achieves operational status.

A mill (3000 tonnes ore per day, 80% process recovery) at the new Turadimh mine is described in the 2005 Red Book as "under construction to treat the ore of the Turamdih and Banduhurang mines." It is further stated that "this plant will undergo expansion at a later date to treat the ore of Mohuldih mine." Additional mills are described as planned at Seripally (1250, 77%), to serve the Lambapur-Peddagattu mine, and at Domiasiat (1370, 87%) to serve the mine located there.

Indigenous Indian uranium production is central to an understanding of the impact of the U.S.-India nuclear accord. The focus in this section has been on the potential (maximum, nominal, capacity) production of India's uranium mines and mills. In the first section of the following chapter, these facilities are again the focal point from the viewpoint of ascertaining the extent to which these potential production capacities have been achieved.

C H A P T E R 4

Fuel Cycle Analysis: From Beginning to Present Day

A complete fuel-cycle assessment of India is presented in this chapter, using a flowsheet description of the relevant materials, facilities and technologies. Results, with explanations for reproducibility of the analysis, are emphasized in this study. The fuel-cycle assessment is based on epochs defined by the significant milestones in the Indian nuclear timeline: 1974 (first nuclear explosion, Pokhran-I test); 1998 (Pokhran-II tests), and 2006 (beginning of pursuit for the U.S.-India Nuclear Cooperation Agreement). The data collected and synthesized for depiction in the flowsheet are representative of the consolidated contribution to the fuel cycle, within periods designated by these milestones. The material accounting, electricity generation, and technologies adopted are shown through the method of timeline depiction in the corresponding flowsheets. The policy approach, with regard to the operation of facilities and their priorities, is understood and interpreted by describing and analyzing the timeline data in sequential topical descriptions of materials produced and consumed. The topical descriptions address the materials produced and consumed and have been arranged in this section on the basis of their relevance in the time frame. The blocks in flowsheets with solely military applications are represented in red, civilian in yellow and the dual-use facilities are shown in blue. The dual-use and inter-mingled status of the strategic and civilian nature of the nuclear program is clearly evident and analyzed in this chapter, from the mass flows represented in the flowsheets.

4.1 ASSESSMENT OF NATURAL URANIUM PRODUCTION

Table 4.1 data represent the evolution in time of various estimates of Indian uranium production. The Red Book estimates were obtained from the tables labeled "Historical Uranium Production" in the various editions of the Red Book (published by IAEA). The other two rows of estimates (capacity-estimate values and working-estimate values) were obtained as follows:

The second row of data in Table 4.1, representing capacity estimates, were obtained under the assumption that the various facilities described in Section 3.7 were operating under the designed capacities and other parametric values indicated in Table 3.9 (for mines or textually in Section 3.7 for mills) with the exception of flows limited by capacities of other facilities in their respective processing chains. For example, the excessive uranium mining may be limited by the available milling capacity, or otherwise, lowered ore production would result in lessened uranium output even if mill is of high capacity. Except as explicitly noted otherwise, most notably for the Jaduguda mine and mill, facility parameters were taken from Table 3.9 of the preceding chapter.

Table 4.1: Estimates of Indian uranium production (in tonnes).

Period	1967–1989	1990	1991	1992	1993	1994	1995	1996	1997
Red Book Estimate	5200	230	200	150*	148*	155*	155	250	207
Capacity Estimate	3984	205	205	205	205	205	222	222	222
Working Estimate	3984	205	205	205	205	205	205	205	205
Period	1998	1999	2000	2001	2002	2003	2004	2005	2006
Red Book Estimate	207	207	207	230	230	230	230	230	230
Capacity Estimate	323	323	323	323	323	355	355	355	355
Working Estimate	180	173	165	132	101	101	101	106	111

* All entries, except these, are labeled in the Red Book as "Secretariat estimate."

In effect, this means five eras were distinguished, for the capacity estimates:

- From 1967 to 1985 capacity production of the Jaduguda mill (assumed to be under operation at 365 days per year) was limited by the 1000 tonnes per (calendar) day of ore (grade .06%, 80% mining recovery), rather than the 1370 tonnes ore per day capacity of the Jaduguda mill (95% maximum process recovery) that serviced the output of that mine.

- Beginning in 1986 the production capacity was increased by the 250 tonnes ore per day (grade .06%, 75% mining recovery) of the Bhatin mine, which still left the mine production capacity below that of the Jaduguda mill.

- From 1995 mine capacity was augmented by the Narwapahar mine (capacity 1000 tonnes ore per day, grade .05%, 80% mining recovery). However, the total mining capacity then exceeds mill capacity, so production from Narwapahar along with Bhatin and Jaduguda was assumed capped at 1370 tonnes per day.

- Beginning in 1998 the capacity of the Jaduguda mill was assumed to be expanded to 2100 tonnes ore per day, so that production at the Narwapahar mine could now be processed (850 tonnes per day (2100-1000-250) only slightly below its capacity of 1000 tonnes per day).

- Finally, beginning 2003, mining capacity was again expanded by the 550 tonnes per day capacity of the Turamdih mine (grade.0025%, 75% mining recovery), which was processed at the Turamdih mill (capacity 3000 tonnes ore per day, 80% process recovery).

Through 1994, the working estimates of Table 4.1 are identical to the capacity estimates just described. However, from 1995 onward the working estimates incorporate *inter alia* the assumption that the last three events in the above list did not happen, by 2006 or earlier. That is, the Narwapahar mine did not produce significant uranium ore in the period 1995-2006, the capacity of the Jaduguda mill did not, by 2006, expand from the prior 1370 tonnes ore per day to 2100, and the Turamdih mine did not begin significant production by 2006. This working estimate can thus be considered as a hypothesis that India has, over the past decade and half, encountered severe and unanticipated difficulties in realizing its planned indigenous production of uranium. The remainder of this chapter then can be viewed as an exploration of the consequences of the hypothesis thus embodied in this working estimate. However, it is appropriate to note that there are ample suggestions in public documents that tend to support this hypothesis, at least to the extent of establishing that severe difficulties were encountered in fully realizing the planned potential of the Narwapahar mine, the Jaduguda mill and the Turamdih mine.[1]

During the period 1967 to 1994, the working estimate was taken as identical to the capacity estimate. From 1967-1990, this averaged approximately 175 tonnes per year, which is substantially less than the corresponding Red Book average of nearly 230 tonnes per year. The basis for the latter is not provided in the Red Books. It closely corresponds to the value one would obtain if production during this period were set at the capacity of the Jaduguda mill (1370 tonnes ore per day), but that would ignore the more-or-less accepted capacity limitations of the Jaduguda mine, which would appear to have been the predominant production bottleneck during this period.

---

[1] For example, in regard to the Narwapahar mine and some instances in regard to the Jaduguda mill: WISE Uranium Project, "Issues at Jaduguda Uranium Mine, Jharkand, India," (accessed April 9, 2009), reports "the capacity of UCIL's uranium ore processing plant at Jaduguda, inaugurated in 1967, was increased to 2,090 tonnes per day from the initial 1,200 tonnes," with an attribution to the Times of India of June 25, 2007. Likewise, WISE Uranium Project, "New Uranium Mining Projects - India" (http://www.wise-uranium.org/upin.html, accessed April 9,2009) reports that "a delay in the commissioning of the milling system at the Jaduguda mines in Jharkhand is a major factor behind the current fuel shortage," with an attribution to the Indian Express of August 20, 2007. J. L. Bhasin, Ashok Mohan (Technical Advisor (IR), DAE) and K. K. Beri (Director (Tech), UCIL) ('Uranium ore in India," August 11, 2006, jansamachar.net, accessed April 9, 2009) state: "UCIL started it's operations in 1968 at Jaduguda with uranium ore mining and a processing plant, each of 1,000 metric tonnes per day (MTPD) capacity. To meet the growing demand of uranium for nuclear power, UCIL has expanded its activities both in mining and processing plants. The focus has been on efficiency, cost effectiveness, as well as on environmental safety. The new mine has been opened at Narwapahar. The processing plant at Jaduguda has been expanded to process this extra ore. Now, the overall processing capacity of the Uranium Mill is 2090 MTPD." Reasons for the possible delay in expansion are by no means clear, but a factor could have been US sanctions placed in the Jaduguda mining and milling complex, in the wake of the 1998 nuclear weapon tests carried out by India. See Bureau of Export Administration, "Dual-use Export Control sanctions: India and Pakistan," November 13, 1998, http://jya.com/bxa-ind-pak2.htm, accessed April 9, 2009. Reported civil unrest in the locale of Jaduguda could also have been a contributing factor. For example, Surendra Gadekar ("India's nuclear fuel shortage," Bulletin of the Atomic Scientists, August 6, 2008, http://www.thebulletin.org/web-edition/features/indias-nuclear-fuel-shortage, accessed April 9, 2009) states: "The more likely reason (for lack of uranium production) is that despite the critical uranium shortage, UCIL officers are unwilling to work in remote areas where the Indian government is facing a Maoist insurgency. A large tract spanning the remote areas of the eastern states of Chhattisgarh, Jharkhand, Orissa, and Andhra Pradesh have serious law-and-order problems." In regard to the Turamdih mine: A news item dated November 22, 2007 ("Commissioning of Turamdih uranium plant in a fortnight," The Hindu, http://www.thehindu.com/2007/11/22/stories/2007112262050800.htm, accessed May 3, 2009) states that: "A new uranium mining plant will be commissioned at Turamdih in East Singhbum district in Jharkhand in a couple of weeks, according to Ramendra Gupta, Chairman of Uranium Corporation of India Limited (UCIL)." This article further specifies that: "…the plant proposed …is held up with the appellate authority on objections filed by environmental groups." Tender Notice No. UCIL/TMD/MECH/(VS) -02/2008, dated December 29, 2008, seems to request bids for equipment essential to operation of the Turamdih mine.

Beginning in 1991, the Red Book estimates show a substantial decline in production of uranium. No incident is indicated as the basis for this decline; however these dates include the few not explicitly labeled as "Secretariat estimates." Nonetheless, given the lack of a basis, this decline is not explicitly incorporated into the working estimates of production shown in Table 4.1. Starting in 1996, the Red Book estimates of Table 4.1 show a recovery from the declining production originating in 1991. The basis for this increase also is unclear. Very possibly, it is associated with the presumed opening of the Narwapahar mine in 1995 (cf. Section 3.7); however, if so, then at least the 1996 estimate appears not to account for the additional bottleneck posed by the capacity of the Jaduguda mill. This mill bottleneck is taken into account in the working estimates of Table 4.1.

From 1997, and continuing through 2002, the working estimates of uranium production in Table 4.1 show a gradual decline. By contrast, the capacity estimates show a sharp increase, while the Red Book estimates remain almost constant, with a slight increase (about 20%) in 2001. Non-incorporation of the capacity increase (of the Jaduguda mill) into the working estimates appears to be the predominant reason for this difference. It is unclear why this capacity increase is not reflected in the Red Book estimates. For the working estimates, it is assumed that expansion has occurred later than 2006, so it plays no role in the working estimates of Table 4.1. The lack of an increase in the capacity of the Jaduguda mill brings the working estimates of Table 4.1 more-or-less in line, with the Red Book estimates, from 1998 onward.

The declining values, from 1997 through 2002, in the working estimates of Table 4.1 stem from the additional premise, based on Red Book data, that production performance at the Jaduguda complex began to decline in 1997 and continued to do so through 2006. Specifically, through 1996 the values listed in the Red Books for ore production at Jaduguda mine and the associated process recovery were, respectively, 1000 tonnes per day and 95%. In 1998, these respective values were 850 tonnes per day and 950% process recovery, and the decline thus initiated continued as follows:

- 2000 (850 t/d, 87% process recovery),

- 2002 (600 t/d, 80% process recovery),

- 2004 (600 t/d, 80% process recovery) and

- 2006 (650 t/d, 80% process recovery).

These values were used, with interpolation for odd years, to calculate the best estimates of Table 4.1.[2]

---

[2] The ultimate causes of these declines in performance are not clear, and there may be multiple causes. A possible contributor to the declining ore production at Jaduguda mine might be the need to go to deeper levels to follow the ore vein, and difficulties encountered in doing so. There are multiple explicit indications of this need, and tacit indications of associated difficulty. In regard to the necessity, Bhasin, Mohan and Beri (op cit.) say: "Jaduguda Mine is the first mine opened by UCIL in 1968. Ore body in Jaduguda has been prospected up to a depth of about 800 m below surface and it is expected that it would continue further in depth. The main entry to the mine is through a circular shaft. The shaft has been sunk to a depth of 640 m in two stages – first stage from surface to 315 m and second stage from 315 m to 640 m. As ore up to the second stage is depleting, UCIL has planned to deepen the mine further." Although this 2006 publication cites the third stage as "planned," there are reports that its construction commenced in December, 1998, and was completed in 2004. (Cf. Ministry of Statistics

Time Frame	Total Uranium Production (tonnes)	Total U_3O_8 Production (tonnes)	Total UF_6 Production (tonnes)[a]	Total UO_2 Production (tonnes)[b]
1967–74	1332	1571	1575	1199
1975–98	4464	5264	5281	4019
1999–2006	991	1169	1173	892

Table 4.2: Working estimates of aggregate indigenous Indian uranium production, by epoch.

[a] A 20% loss of uranium is estimated to occur in the conversion from U_3O_8 to UF_6.
[b] Incorporates 0.8% fabrication loss.

Table 4.2 displays the corresponding working estimates of indigenous Indian uranium, by the time epochs described in the preface to this chapter. These estimates were obtained in the same manner as the working estimates of Table 4.1, as described earlier in this chapter. They will provide the basis, on the production side, for the analysis of the historical flow of materials in the Indian nuclear fuel cycle that will be developed in the remainder of this chapter.

This uranium emerges from the mining and milling process in the form of U_3O_8, commonly termed as "yellowcake." The corresponding amounts of this compound estimated to have been produced during the three epochs are shown in the third column of Table 4.2. UF_6 is stable, so is a suitable form for long-term storage. It is converted to UO_2 for fabrication into reactor fuel, either directly from natural uranium (0.071% U-235) content, or after enrichment. UF_6 is a gaseous material suitable as feedstock for the enrichment process. Equivalent amounts of UF_6 and UO_2 are shown in the respective last two columns of Table 4.2. The UO_2 quantity was calculated assuming all uranium was retained in the natural isotopic ratios, which is the suitable composition of fuel for the existing CANDU type reactors in India.

4.2 ASSESSMENT OF POWER PRODUCTION AND URANIUM CONSUMED

India's nuclear power plant analysis involves computing the quantity of fuel used along with the accumulation of the spent fuel. The capacity factors listed in Tables 3.2 and 3.3 (Section 3.1) were used to compute the UO_2 consumed by each facility. Initial CANDU core loadings do not consist of exclusively fresh UO_2 fuel bundles, so as to avoid flux peaking due to plutonium production leading to lowered delayed neutron fraction. Rather, selective positioning of depleted uranium fuel bundles flattens the flux and improves controllability during fuel reshuffling stage.[3]

and Programme Implementation (Infrastructure and Project Monitoring Division), "Infrastructure projects completed in the year 2004-05," `http://www.mospi.gov.in/proejct_brief_for_MOS_press_conference_allindia.htm`, accessed April 7, 2009.) The decrease in process recovery could have been occasioned by the factors affecting the planned mill expansion, as discussed in the preceding footnote.

[3] B. Rouben, CANDU Fuel-Management Course, Atomic Energy of Canada Limited, Montreal, Canada (1999).

The fuel quantities, refueling sequence and reshuffling period have all been studied with detailed computations via the two-dimensional current coupling physics code (HELIOS), on 37-pincell CANDU fuel bundle design. The results of this analysis are displayed in Tables 4.2, 4.3, and 4.4. For the 220 MWe and 540 MWe rated power reactors, the initial core loading is of 57.65 tonnes and 141.5 tonnes of UO_2, respectively. The first criticality of the core can be achieved either with a combination of fresh and depleted uranium fuel or fresh uranium fuel with a few bundles of thorium. RAPS and MAPS received their initial loads of 17.65 tonnes of spent fuel from CIRUS reactor. Spent fuel discharged from the RAPS and the MAPS was used for the initial fuel loading at the Narora Power Station (NAPS). But from Kakrapar Atomic Power Station (KAPS-2) onwards, thorium bundles were used along with fresh UO_2 fuel bundles for the first core loading.

After first criticality is attained, the k-effective of the reactor drops from 1.06 due to burning of uranium. It then starts to climb up with the production of plutonium from U-238. It eventually declines to the level where it requires a first increment of fresh fuel after energy generation equivalent to 140 days of operation at full power. These first 140 full power days can be termed as pre-refueling period. The excess reactivity to meet the power demand and keep the reactor critical is met by plutonium production and the fresh uranium present in the fuel bundles. The core flux map though needs flux flattening because of spatially varying production of fissile material, which is addressed by reshuffling of fuel bundles. During this pre-refueling period, there is a saving of 17.6 tonnes of uranium in 220 MWe reactors and 43.2 tonnes in 540 MWe reactors because of the non-refueling days.

These estimates of UO_2 consumption were cross checked[4] by computing the corresponding burn-ups. All resulting values were in the range 6500-7500 MW_{th}*days/tonne of Uranium, as expected for CANDU-type reactors.

The total amount of UO_2 consumed until 2006 is determined through the sum of all quantities in Table 4.3 and 4.4 along with the corresponding fresh fuel loads. The initial fresh fuel load was 40 tonnes each for the 12 plants with 220 MWe power rating and 98.3 tonnes for TAPP-3, 4. Thus, the total amount of UO_2 consumed is 5007 tonnes until the end of 2006. The total amount of UO_2 that was produced until then was 6110 tonnes (sum of the last column in Table 4.2).

In addition as will be shown in Sections 4.4 through 4.6 below, the production (research) reactors (CIRUS and DHRUVA), consumed 609 tonnes of UO_2 by the end of 2006. Thus, the total UO_2 reserve available at the end of 2006 was 494 (6110-5007-609) tonnes. Assuming no additional mining activity is added and no amount of UF_6 was diverted for enrichment the reserves and production for the power reactors can last for only a few more years. As detailed later in the flowchart descriptions, 25.75 tonnes of UF_6 was used for U-235 enrichment. This quantity of UF_6 would bring down the reserves of UO_2 by 20 tonnes making it 474 tonnes by the end of 2006.

---

[4] Efficiencies used were 30% for RAPS-1,2, MAPS-1,2, NAPS-1,2 and KAPS-1. Account was taken of the use of depleted uranium from the research reactor in the initial loading of the power reactors. (This quantity of depleted uranium fuel is charged against use in the following section of plutonium production.) The value 27.5% was used for the efficiency of the remaining five reactors in Table 3.2. These reactors used thorium, rather than depleted uranium in the fresh core, in order to operate at the rated power during the plutonium phase of initial days of reactor operation.

Table 4.3: Fuel consumed by PHWRs until 2003.

Plant	UO$_2$ Used (tonnes)
RAPS-1	255
RAPS-2	436
MAPS-1	378
MAPS-2	339
NAPS-1	274
NAPS-2	281
KAPS-1	267
KAPS-2	254
KAIGA-1	91
KAIGA-2	122
RAPS-3	88
RAPS-4	90

Table 4.4: Fuel consumed by PHWRs from 2004 to 2006.

Plant	UO$_2$ Used(tonnes)	Year
All 12 Plants from Table 3.2	366	2004
TAPP-4 and All 12 Plants	352	2005
TAPP-3, TAPP-4, and all 12 Plants	257	2006

Table 4.5 shows the fuel consumption for 2007 and 2008 from present operating plants and newer additions. During that period, all the NPP's in India operated at 60% or lower capacity factor. The same is the case for the newly constructed power plants that lined up at the declared dates.

Summing up the consumption for 2007, as stated in Table 4.5, shows that by Dec 2007, India would have consumed an additional 392 tonnes of UO$_2$. If the operating capacity factors had remained at 60% for 2008, then according to last column of Table 4.5, an additional 439 tonnes of UO$_2$ would have been consumed in the period from Jan 2008 to Dec 2008. Thus, 831 tonnes (392+439) of UO$_2$ would have been consumed in the period from Jan 2007 to Dec 2008. The production of UO$_2$ in 2007 and 2008 is calculated (by extrapolation from the prior working estimates) to be 116 and 121 tonnes, respectively. So the total UO$_2$ available from the reserve of 2006 and production in 2007 and 2008 is 711 tonnes (=474+116+121). Thus, the consumption exceeds the production for an assumed capacity factor of 60% for 2008. The consumption would

Table 4.5: UO_2 fuel consumption by PHWRs at 60% capacity factor in 2007 and 2008.

Plant	2007 Consumption (tonnes)	2008 Consumption (tonnes)
TAPP-4	55	55
TAPP-3	55	55
KAIGA-3	17	23
KAIGA-4	6	23
RAPP-5	6	23
RAPP-6	0	7
RAPS-2	23	23
MAPS-1	23	23
MAPS-2	23	23
NAPS-1	23	23
NAPS-2	23	23
KAPS-1	23	23
KAPS-2	23	23
KAIGA-1	23	23
KAIGA-2	23	23
RAPS-3	23	23
RAPS-4	23	23

have exceeded the production and reserve by flag end of 2008 even if operational capacity factors were further dropped to 50%.

4.3   ASSESSMENT OF ENRICHED URANIUM PRODUCTION

Modeling of India's enrichment facilities (Subsection 4.3.1) shows that India could have accumulated 20 kilograms of 90% U-235 by 1998 and 94 kilograms of 90% U-235 by the end of 2006 (assuming 10 kilograms used in the Pokhran-II tests and 16 kilograms used as experimental fuel in DHRUVA reactor). This amount of enriched uranium can fuel a nuclear submarine core (Subsection 4.3.2), if India continues that program.

It is also shown, in Subsection 4.3.1, that the Mysore enrichment plant needs a feed of 2.15 tonnes of UF_6 per year for producing 10 kilograms of 90% U-235. The Trombay plant uses a feed of 0.43 tonnes of UF_6 every year to produce 2 kilograms of 90% U-235. The calculations of the feed capacity for the enrichment plants, and production details on an annual basis, are also developed in the following subsection. The resulting enriched uranium production with the actual feed requirements, SWU capacity, and centrifuges of a particular design are shown in Figures 4.1 and 4.2. These are based on the assumption of operation at full capacity over an entire year.

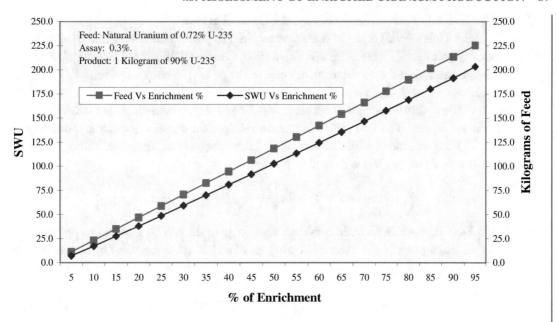

Figure 4.1: Feed and SWU capacity to attain the percentages of enrichment.

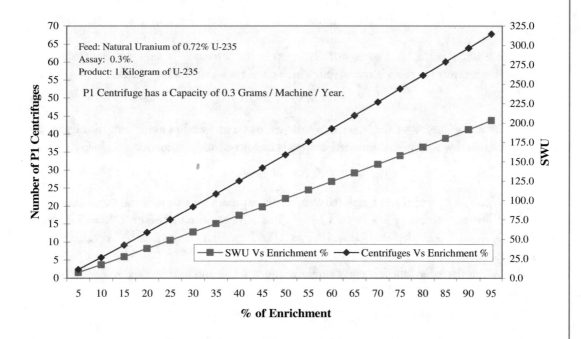

Figure 4.2: SWU capacity and centrifuges required to attain the percentages of enrichment.

4.3.1   THEORY OF ENRICHMENT CALCULATIONS

Separative Work Unit (SWU) describe the effort required to produce in the various types of separation devices a unit mass (Electromagnetic Separation, Gaseous-Diffusion, Centrifugal, Aerodynamic, Laser, Chemical and etc.) for relative enrichment of U-235 to the amount of U-238. As shown in Figures 4.1, 4.2, and justified below approximately 192 SWUs is required to produce 1 kilogram of 90% U-235 from natural uranium of 0.72% U-235, with a residue of 0.3% U-235 as depleted uranium waste.[5] The feed "F" in kilograms required for a desired quality and quantity of product "P" in kilograms along with the waste "W" (kilograms) was calculated following Mozley:[6]

The value function $[V(N)]$ is defined as follows

$$V(N) = (2N - 1) \times \ln\left[\frac{N}{1 - N}\right] + b \times N + a, \tag{1}$$

often on the basis of thermo-dynamic considerations. It gives the SWUs required to produce one kilogram of enriched product N_p. Here the constants 'a' and 'b' are determined by setting to zero the value function and its derivative with respect to N at $N_0 =$ value of N in the feed. This gives:

$$a = - \ln\left[\frac{1 - N_0}{N_0}\right] - \left[\frac{1 - 2N_0}{1 - N_0}\right], \tag{2}$$

$$b = 2 \times \ln\left[\frac{1 - N_0}{N_0}\right] + \left[\frac{1 - 2N_0}{N_0 \times (1 - N_0)}\right]. \tag{3}$$

The value of N_0 is 0.0072 for natural uranium. Thus, for natural uranium $a = -5.933$ and $b = 149.72$.

Since material is neither created nor destroyed in the separators, then for each separator or collection of separators the mass is conserved, which gives the conservation equation

$$P + W = F \tag{5}$$

Where, P, W and F are, respectively, mass of product, waste and feed per unit time. Similarly, the sum total fractional weight of the enriched isotope is conserved and the equation is given by

$$N_p * P + N_w * W = N_f * F \tag{6}$$

Where the N_p, N_w, N_f and N_0 are fractional weights of the product, waste, feed and natural uranium, respectively. For production of 1 kilogram ($P = 1$) of 90% enriched uranium ($N_p = 0.9$) (90% enriched U), $N_w = 0.003$ (depleted U) and $N_f = 0.0072$ (natural U) the feed "F" is calculated to be 213.57 kilograms of natural uranium (0.72% U-235).

Evaluating the value function equation for product, waste and feed terms the required separative work units can be obtained by the following equation,

$$SWU = V(N_p) * P + V(N_w) * W - V(N_f) * F \tag{7}$$

[5] M. V. Ramana, "India's Uranium Enrichment Program," INESAP Information Bulletin 24 (Dec 2004).
[6] R. F. Mozley, The Politics and Technology of Nuclear Proliferation, University of Washington Press, Seattle, 1998.

It is calculated from Equation (7) that 192 SWU is required to achieve the desired product, at the specified tail assay.

4.3.2   THE NUCLEAR SUBMARINE PROGRAM

Little is known about the nuclear submarine design and the number of them being planned for strategic deployment. Going by the standard designs, it is assumed that India would need 100 kilograms of 90 w/o U-235 for each submarine core. The basis of the assumption is explicitly calculated (w/o = weight percent, a/o = atomic percent).

The standard design speed of a nuclear submarine would be around 30–35 knots (30 knots = 15 m/sec). Estimates of the U-235 (100% enriched U) for U.S. nuclear submarines is close to 0.6–0.7 grams per shaft horse power (shp) per year. The requirements of Russian submarines are likely to be about 0.315-0.35 grams/shaft horse power/year. The difference is because fast reactors with high fuel enrichment are used by the Russian submarines and PWR cores by the U.S. submarines. The propulsion power for Charlie Class Submarine is 20,000 shp. Due to the smaller distances that the ATV (Advanced Technology Vessel, name coined for the nuclear submarine of India) is likely to traverse, it could be assumed that ATV would require about 0.3 grams/shaft horse power/year (1 shp = 746 watts).

Considering propulsion power of 20,000 shp and refueling requirement of once in 15 years, the ATV requirement for U-235 = 0.3 grams/shp/year * 20,000 shp * 15 years = 90 kilograms U-235. That is equal to 100 kilograms of 90 w/o U-235.

The uranium enrichment activity at the Mysore plant began in 1990 with a higher end speculated installed capacity of 2000 SWU/year. Thus, over a period of 10 years, the facility would have produced 20,000 SWU. It requires 192 SWU and 213 kilograms of natural uranium to produce 1 kilogram of 90% enriched U-235. Therefore, the total expected 90% enriched U-235 accumulation over 10 years period is 104.4 kilograms. It is approximately the quantity of material required for one core of fuel for the proposed Indian nuclear submarine. Thus, the fuel for the first nuclear submarine of India, as well as an additional reactor core every 10 years, would be available irrespective of the status of U.S.-India nuclear accord. The production of uranium for future submarine reactor cores at a faster rate however may require additional resources.

4.4   ASSESSMENT OF PLUTONIUM PRODUCTION AND USE: 1947–1974

Plutonium production in India began with the CIRUS reactor, the history of which has been previously discussed in detail. In the flowsheet assessment until 1974 (Figure 4.3), the CIRUS reactor fuel was analyzed using the buildup and depletion code ORIGEN2.2.[7] Simulations were performed assuming a specific power of 32 watts/gram of U, with 15-day burn-up steps, and using a natural uranium CANDU reactor cross section library. The CIRUS reactor was the source of

---

[7] A.G. Croff, "ORIGEN2.2: Isotopic Generation and Depletion Code Matrix Multiplication Method," ORNL/TM-7175, ORNL (July 1980).

depleted uranium fuel for RAPS-I reactor and also weapons-grade plutonium for the first nuclear weapon test of India. Both the commitment to peaceful use and intention for weapons development was clearly evident, by the dual use nature of the nuclear technology that India pursued.

The 220 MWe rated RAPS-I reactor was originally made critical using 40 tonnes of natural uranium and 17.65 tonnes of depleted uranium fuel (at a burn-up of 1066 MWday/tU) obtained from CIRUS reactor. (See lower left of Figure 4.3.) The spent fuel of CIRUS was reprocessed in the PHOENIX plant (mid-upper right of Figure 4.3). The spent fuel reprocessing for use in the RAPS-I core only involved fission product removal. The use of the depleted uranium instead of all fresh natural uranium fuel was to diminish the effect of the plutonium peak typically seen in CANDU reactors when burning all fresh fuel. After the first criticality of the CANDU reactor with natural uranium fuel, plutonium production from U-238 begins. Originally, plutonium is produced at a higher rate than it is consumed by fission. This excessive plutonium accumulation decreases the delayed neutron fraction of the fuel, which necessitates faster acting regulation system and lowers the reactor safety margins. To cater to the lowered action time available for the reactor regulation system in case of a transient, the core power is reduced until plutonium levels stabilize. However, by having depleted uranium bundles in selective channels the fissile content is reduced and the reactor regulation system can control the core power transients with the reactor operating at full power. The quantity of the depleted uranium required was computed using the 2D lattice physics code HELIOS-1.4.[8] In the computation of the quantity of depleted uranium fuel required, a basis of 92 mk excess reactivity was assumed. Excess reactivity less than that of 92 mk leads to a higher use of depleted uranium fuel in the core. If Bhabha's plan was to be realized, then savings in depleted uranium was a necessity. This plan required that in 12 years (1974-1986), 9 power plants of 220 MWe rated were supposed to be built. The depleted uranium requirement for these reactors at the rate of 17.65 tonnes would have been 158.85 tonnes. This need could have been barely meet with reserves and annual production along with supplying the weapons initiative. If the excess reactivity was to be greater than 92 mk, then more absorber rods would have to be installed challenging the compact core size of the CANDU reactors. The neutron absorption could also have been achieved with liquid poison, but that would necessitate higher reshuffling sequences to diminish the plutonium peak. The limit of the refueling machine cannot be ignored under the circumstances. These assumptions and conclusions have been reached on the basis of reactor operations experience, knowledge of reactor theory and understanding of the design details.

To calculate the quantity of plutonium produced in the time period, an estimate is made of the amount of fuel irradiated. Usually the amount of fuel irradiated is based on the figures published for the amount of electricity generated in the case of power reactors. Since electricity production is not of relevance in these reactors, a capacity factor of 50% is calculated[9] for CIRUS to compute

---

[8] E.A. Villarino et. al, "HELIOS: Angular Dependent Collision Probabilities," Nuclear Science and Technology, 112, p. 16-31 (1992).

[9] Capacity factor calculations for the CIRUS reactor: Step (1) Burn-up required to achieve weapons-grade plutonium quality = 1066.67 MWd/tU. Step (2) This is equivalent to (1066.67 / 365.25) MW-yrs/tU = 2.92 MW-yrs/tU. Step (3) Total quantity of fuel used by CIRUS = 10.5 tonnes of UO_2 = 10.5 * (238/270) tU = 9.25 tU. Step (4) the total heat produced from 9.25 tU

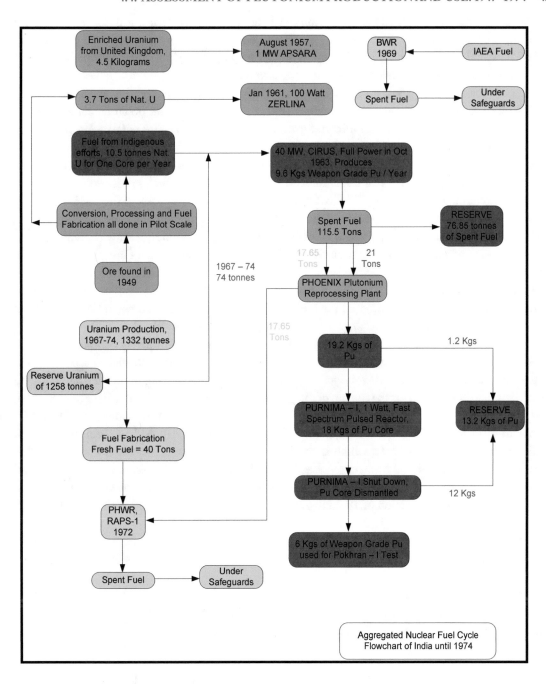

Figure 4.3: Nuclear fuel cycle flowsheet until 1974.

plutonium estimates. Weapons-grade plutonium is defined as that with a ratio of 6% or less for Pu-240/Pu-239. The lower percentage is considered because the spontaneous fission capability of Pu-240 would tend to set off a chain reaction prematurely resulting in a fizzle, with perhaps only a few tonnes of energy being released as against its nominal value.

CIRUS became critical on the 10th of July 1960, but it started operating at full power only in October of 1963. By the end of September 1997, the reactor was shut down for refurbishment. In October 2000, it was still undergoing refurbishment. It is an old reactor with pneumatic control systems and requires an extensive period for refueling. As concluded from calculations with an annual capacity factor of 50%, the maximum can be reached in a year at full power operation of the core. From the output of ORIGEN2.2 depletion code for CIRUS reactor fuel, production of weapons-grade plutonium was computed at 280 days at full power, which corresponds to the desired burn-up. This allows for 85 days in a year for shutdown to do maintenance, any refueling and make-up for lost hours of low capacity factor operation. CIRUS can irradiate 10.5 tonnes of natural uranium oxide fuel per year,[10] which results in accumulation of 115.5 tonnes (center right of Figure 4.3) of depleted uranium in the 11-year period 1963-1974. Thus, annual production of weapons-grade plutonium would be 9.6 kilograms, with due allowance for unplanned shutdowns and regular maintenance. In order to be conservative on estimates, the fuel irradiation during the period when CIRUS had just been commissioned and operated at very low capacity factor was ignored.

As indicated in Section 2.1, the PHOENIX reprocessing plant officially became available in 1964, with the designed ability to reprocess 20 tonnes of spent fuel per year, and to separate up to 10 kg of plutonium annually. Its product consisted of: plutonium-bearing depleted uranium, with higher actinides removed, for the startup of the power reactors; and plutonium, with higher actinides, for use in reactors or weapons. However, a variety of well-documented difficulties at PHOENIX prior to 1974 put limitations on the quantity of the spent fuel that could have been reprocessed. The operating assumption here, which is consistent with numerous reports, is that prior to 1974 the PHOENIX plant would have reprocessed 21 tonnes of spent fuel for plutonium extraction and another 17.65 tonnes of spent fuel required for the fresh core of RAPS-1. The spent fuel input of 21 tonnes results in production of 19.2 kilograms of plutonium as computed with the CIRUS fuel ORIGEN code model.

The bulk of the weapon-grade plutonium obtained from this reprocessing (18 kgs out of 19.2 kgs) was used in the fast spectrum PURNIMA reactor for nuclear physics studies. The only option left to make plutonium available for a nuclear device to be tested was by dismantling of the PURNIMA core. Considering that 6 kilograms of the total plutonium was used for the test device, a meager amount of 13.2 kilograms of plutonium was in stockpile for the post-test period. The corresponding flows are depicted in the lower-right portion of Figure 4.3.

---

is 9.25 tU $*2.92$ MW-yrs/tU $= 27.03$ MW-yrs. Step (5) for capacity factor over a year with 280 days of operation at 40 MW_{th} is (27.03 MW-yrs / 40 MW_{th}) * (280 days/365.25 days) = 0.5179 ~ 50%.

[10] http://www.barc.ernet.in/webpages/reactors/cirus.html, "CIRUS Reactor," Bhabha Atomic Research Center, Mumbai (August 2007).

For completeness, three relatively minor usages of uranium are depicted along the top layer of Figure 4.3. Some of these have had substantial impact on the Indian nuclear program.[11] However, the net impact of these activities upon India's supply of fissile material is minor, so they are not further discussed here.

4.5   ASSESSMENT OF PLUTONIUM PRODUCTION AND USE: 1947–1998

Following the pattern of study in Section 4.4, DHRUVA's annual plutonium production capacity was determined using ORIGEN2.2 fuel burn-up and depletion code. On the lines of the CIRUS reactors parametric assumptions DHRUVA reactor's fuel was taken as irradiated at a specific power of 32 watts/gram of U until burn-up of 1055 MWdays/tU was reached, which is required for the plutonium in the spent fuel to qualify as weapons-grade plutonium. It was found that DHRUVA has a much shorter cycle (67 days) for in-core fuel to produce weapons-grade plutonium. Under a pragmatic assumption of five core changes per year, DHRUVA can produce 27.63 kilograms of weapons-grade plutonium annually.

DHRUVA became critical in August 1985, but had various operating difficulties, inclusive of vibrations in the fuel elements and oil leakage from the coolant pump. In December 1986, it began operating at 25 MW_{th}, increasing gradually to reach 100 MW_{th} in January 1988. Since then it is estimated that DHRUVA was operating at an average capacity factor of 75%.[12] Unit start-up period with erratic operating schedule and at very low capacity factor was neglected. DHRUVA has a much shorter cycle than CIRUS, thus requires multiple core changes in a year. As computed by the ORIGEN2.2 code DHRUVA need to operate for 67 days, with 6.35 tonnes of fuel per core change, to attain an average burn-up of 1055 MWday/tU at a specific power of 32 watt/gram of Uranium. It was thus calculated that DHRUVA produces 5.53 kilograms of weapons-grade plutonium per operating cycle. By assuming five core changes per year, DHRUVA can produce 27.63 (= 5 ∗ 5.53) kilograms of weapons-grade plutonium per year.

From the plutonium production capabilities of CIRUS and DHRUVA as described above in this section, it is estimated that the total plutonium production of India by the end of 1998 would have been 393 kilograms (Table 4.6) after taking into account the losses in reprocessing, actual years of operation and the estimated capacity factor of the operating reactors. Assuming 6 kilograms of weapons-grade plutonium was used for Pokhran-I tests, 24 kilograms of weapons-grade plutonium was used for Pokhran-II tests, and 50 kilograms for FBTR core, India would have had enough weapons-grade plutonium (313 kilograms, or approximately 39 IAEA significant quantities) by the

---

[11] For example, arranging a supply of IAEA-safeguarded fuel for the US-built Tarapur 1 and Tarapur-2 reactors has been a continual bone of contention between the US and India, for over thirty years.

[12] Capacity factor calculations for the DHRUVA reactor: Step (1) Burn-up required to achieve weapons-grade plutonium quality = 1055 MWd/tU. Step (2) This is equivalent to (1055 / 365.25) MW-yrs/tU = 2.89 MW-yrs/tU. Step (3) Total quantity of fuel used by DHRUVA = 6.35 tonnes of UO_2 in one core change = 6.35 ∗ (238/270) tU = 5.597 tU, for 5 core changes the fuel quantity = 27.98 tU. Step (4) the total heat produced from 27.98 tU is 27.98 tU ∗2.89 MW-yrs/tU = 80.88 MW-yrs. Step (5) for capacity factor over a year with ((5 core changes / year) ∗ (67 days of operation / core change)) at 100 MW_{th} is (80.88 MW-yrs / 100 MW_{th}) ∗ (5 ∗ 67 days/ 365.25 days) = 0.7418 ∼ 75%.

Table 4.6: Plutonium produced and uranium used in CIRUS and DHRUVA.

Timeframe	Quantity of Weapons-Grade Plutonium Produced (kg)	Quantity of Natural Uranium Irradiated (tonnes)	
1964–1974	19.2	CIRUS	20.99
		DHRUVA	Not Applicable
1964–1998	393.31	CIRUS	173.25
		DHRUVA	269.875
1964–2006	633.50	CIRUS	204.75
		DHRUVA	485.775
2006–2011(Projected)	140.93	DHRUVA	107.95

year 1998 (center bottom in Figure 4.4) for a formidable deterrent of implosion devices in Southeast Asia.

In Table 4.6, time frames ending in 1974, 1997 and 2006 were chosen because of first implosion device test in 1974, one year prior to Pokhran-II tests and nuclear fuel cycle assessment and cooperation agreement initiation being 2006. A projection until 2011 is shown for plutonium accumulation, if CIRUS is decommissioned as of now and DHRUVA continues to operate at a capacity factor of 75%.

Using the facility and material data from Chapter 3, the spent fuel accumulated along with the fuel's characteristics computed using ORIGEN2.2 and HELIOS-1.4 lattice code for 37-pincell CANDU fuel bundle was analyzed. Multiplication factor calculations over burn-up, fuel isotopics, cross-sections and delayed neutron fraction was executed with one, two and 34 energy group neutrons. An adjusted cross-section library of 34 neutron and 18 photon energy groups was universally used over all the input decks.

The fuel characteristics were computed for fuelmaps of a pincell, a fuel bundle and for the complete assembly comprising of pressure tube (PT), calandria tube (CT), moderator and coolant in the control volume. The fuel isotopics data for two energy groups, with cut-off at 0.623 eV, had acceptable deviations from values for the full 34 energy group data simulations. The final isotopics presented was for the assembly fuelmap comprising of structural materials, moderator and coolant.

The HELIOS input module AURORA developed with the design details of CANDU fuel bundle was executed and the results were generated by the ZENITH output module. The details of the plutonium produced and uranium used are specified in Table 5.3 of Chapter 5. A projection for future use of resources and policy decisions is provided in the next chapter, on the basis of the extensive HELIOS CANDU fuel bundle simulations.

The usage of resources until 1998 was calculated on the basis of the available information on energy generation by India's nuclear power plants. A cross-check of the capacity factors stated in Table 3.2 and 3.3 was done taking into account energy generation and plant efficiency. Stated in the left-center of Figure 4.4 is the break-up for the quantities of UO_2 used by then existing eight CANDU power plants. As can be observed, the latest units built before 1998 consumed more fuel

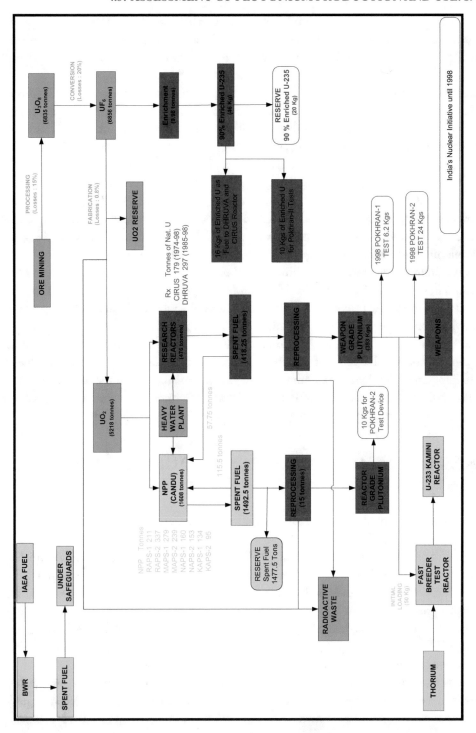

Figure 4.4: Nuclear fuel cycle flowsheet until 1998.

over the same period of operation than the older plants. This is because of the higher capacity factor of operations due to standardization of 220 MWe CANDU reactors by India. This set a prominent goal for building more reactors but the larger picture of uranium availability for future reactors still stood unresolved.

This study considers 1998 as a vital juncture in India's nuclear endeavor because of the Pokhran-II tests and start of the sharp decline of the domestic uranium resources. The nuclear era before 1998 was marred by lack of international technological support. Post 1998, an additional burden of diminishing uranium resources made the fuel availability even bleaker for the existing plants. These events set the tone for international civilian nuclear collaboration in the early years of this decade. Nonetheless, technological advances for implementing Bhabha's three stage nuclear power program never ceased. By 1998, two more new entrants the FBTR and Kalpakkam Mini (KAMINI) reactor were also demonstrated. As stated earlier, FBTR laid the foundation for studies and experiments on fast breeders and paved the path for construction of PFBR. Pursuance of the U-233 reactor KAMINI would materialize the third stage of the Bhabha's nuclear power program.

Given the limited reprocessing capabilities of India, it is assumed that most of the spent fuel from PHWRs is in storage for future use. India's second-stage power generation program demands plutonium for the fast breeder reactors. The plausible course of action India may take to generate energy under the three-stage program and maintain a deterrent is analyzed in the next chapter.

At the right-center of the Figure 4.4, amount of UF_6 used for enrichment is indicated. Given the facility capabilities, 46 kilograms of enriched uranium is calculated to have been produced. Assuming 16 kilograms of enriched uranium was used as experimental fuel in the research reactors and 10 kilograms for the test device of the Pokhran-II tests, one finds India would have been left with 20 kilograms of reserve for possible use (partially) in a nuclear submarine core.

4.6   ASSESSMENT OF PLUTONIUM PRODUCTION AND USE: 1947–2006

A consolidated fuel cycle flowsheet assessment for India was similarly performed and analyzed in this section, from beginning of fuel cycle until 2006, using facility information from Chapter 3. The details of the assessment are shown in Figure 4.5. By December 2006 (center bottom in Figure 4.5), it is estimated that India would have accumulated 633.5 kilograms of weapon-grade plutonium. The contribution of CIRUS is estimated to be 187.23 kilograms and from DHRUVA to be 446.28 kilograms of weapon-grade plutonium.

Weapon-grade plutonium contains mainly Pu-239 with a half-life of 24,400 years and a very small quantity of Pu-241, which is a highly undesirable isotope. Pu-241 decays to americium-241 which is an intense emitter of alpha particles, X-rays and gamma rays. Plutonium-241 has a half-life of 13.2 years which means Am-241 accumulates quickly causing serious handling problems. Because of this short life of weapons-grade plutonium due to americium poisoning, the use of weapons-grade plutonium for weapons use is ensured by mixing the freshly produced plutonium with the pre-existing

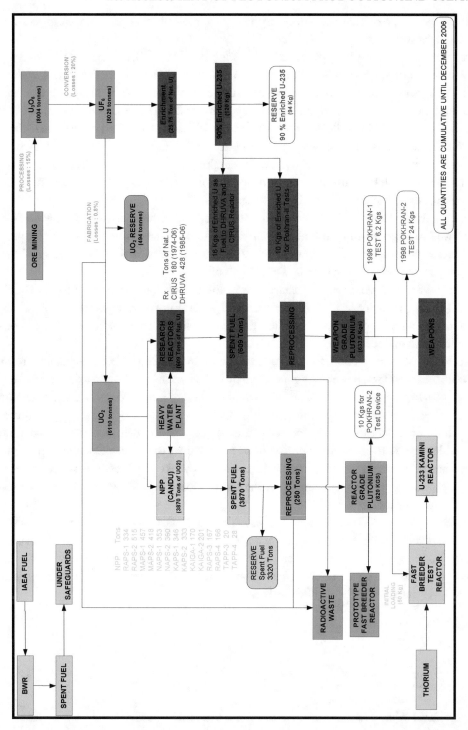

Figure 4.5: Nuclear fuel cycle flowsheet until 2006.

quantities. In either case, replacing the older weapons-grade plutonium or adding to it gives India the nuclear deterrent it seeks.

The reprocessing plant capacity of 250 tHM/yr (stated in Table 3.8), operational for five years, is substantial for separating reactor-grade plutonium, to be used as driver fuel for the PFBR, with an estimated 850 kilograms of plutonium being required to fuel each core. With the assumption that the reprocessing plants are operational at full capacity factor since 2002, it can be concluded that by 2012 the plutonium would also be available for another PFBR as and when constructed. Thus, at this rate of operation of the reprocessing plant and the availability of spent fuel, a fast breeder reactor (FBR) core can be added every five years, without taking the operational FBR doubling time into consideration, i.e., with the plutonium derived largely from PHWRs. A further course of action feasible in realizing more fast-breeder reactors is discussed in Chapter 5.

Continuing the calculations and assumptions on uranium enrichment described in Section 4.3, India would have produced 94 kilograms of 90% enriched U. After accounting for its use as experimental fuel in research reactors and Pokhran-II tests, India was close to having the enriched U fuel for the first core of the nuclear submarine by second half of 2007.

Between 1998 and 2006, there was sharp rise in the use of uranium fuel by the CANDU reactors due to higher capacity factors of operation and addition of newer plants. 2006 and the few years before also witnessed a drop in uranium production due to drying up of mines and non-availability of international fuel.

Similar to the previous section, the usage of resources until 2006 has been calculated on the basis of the available information on energy generation by India's nuclear power plants. A cross-check of the capacity factors stated in Table 3.2 and 3.3 was done taking into account energy generation and plant efficiency. Stated in the left-center of Figure 4.5 is the break-up for the quantities of UO_2 used by then existing 14 CANDU power plants. As can be observed, the latest units built have consumed lesser fuel over the same period of operation than the older plants because of the low capacity factor of operation of the former.[13] In many ways like the 1998 milestone, the year 2006 also set a prominent goal for opening up of the India's closed nuclear sector for international safeguards and collaboration.

This study considers 2006 as the year that brought the de-facto nuclear state to the nuclear and non-weapons states with international safeguards on its facilities. The second stage of Bhabha's three-stage nuclear power program was reaching the point of possibility with the advanced state and progress of the PFBR project. The challenge of having more fast-breeder reactors with thorium cycle is a step closer but far from being effectively realized.

---

[13] The low capacity factor of the newer units seems predominantly to stem from the new 540 MWe TAPP-3 and -4 units.

CHAPTER 5

Fuel Cycle Analysis: Future Projections

In this chapter, the future of the Indian nuclear fuel cycle is assessed in the context of the options available to a state with significant capability in nuclear technology, but receding uranium resources (the "working estimate" of the preceding chapter) and perceiving itself as potentially vulnerable to the threat of two nuclear-armed neighbors. Emphasis is placed upon the effect of the U.S.–India nuclear accord ending a sustained long-standing international isolation following the 1974 Pokhran-I implosion weapon test. The projections of facility utilization and material distribution were done utilizing the resources and capabilities reported in previous chapters. The future of the existing nuclear power program, inclusive of planned breeder program and possible weapons initiatives, is studied with options that may evolve under given circumstances. Details of the U.S.–India nuclear cooperation agreement and its implications on the planned power and future of weapon program are also analyzed. As earlier, all the calculations were conducted with simulations on input decks for 37-pincell CANDU fuel bundle in the two-dimensional lattice code HELIOS-1.4 using 34 neutron and 18 photon energy groups. Complete fuel assembly inclusive of fuel, clad, annular gas between moderator tube and coolant tube, coolant and moderator in one pitch length and width area was considered as a fuel map. Fuel characteristics were reported for a combination of 67 elements and isotopes, in discrete energy spectra.

5.1 PRESUMING NO AGREEABLE U.S.–INDIA NUCLEAR COOPERATION

This section is devoted to contemplation of the near future for India's nuclear program, under circumstances of the previous working estimate of indigenous uranium production and absence of the recently concluded U.S.-India accord on civil nuclear energy. In the first subsection, it is argued that the absence of this agreement, and the resultant access to the international market for uranium (as well as other aspects of modern nuclear technology), would have been devastating to India's current and pending fleet of reactors for purposes of providing significant civil energy. In the second subsection, it is shown that the already existing supply of plutonium from India's fleet of research reactors and PHWRs could, subject to adequate reprocessing capacity and solution of significant technological challenges, support a significant program for development of fast breeder reactors, and the potential for using these breeders for civil purposes is explored. The third subsection is concerned with alternatives related to the potential contribution to a nuclear arsenal of either India's existing or pending fleet of reactors, predominantly PHWRs, or of its pending fast reactors. A significant conclusion is that India could have had, absent the U.S.-India agreement, significant potential for production of nuclear weapons, and that this could have been a significant competitor to peaceful uses of those resources.

5.1.1   FUTURE OF EXISTING NUCLEAR POWER PROGRAM

Under the working estimate of Section 4.1, it was concluded in Section 4.2 that the uranium ore reserves, in tandem with annual production, reached a null point by December of 2008. This date was obtained by calculating the actual uranium consumed (as stated in Tables 4.3–4.5) by power plants and research reactors, and compared against past and present production levels of uranium. The projections of uranium consumption and operating-plant capacity factors for the future were made on the basis of the pragmatic approach a state likely would take, when foreseeing an end to its uranium resources.

From December of 2008, if one assumes India continues to have an annual production of 120 tonnes of UO_2 (extrapolated value on the basis of recent years production rate; cf. Table 4.1), it cannot meet the needs of its nuclear power plants, which already contribute less than 3% of the electricity needs.[1] It is argued, subsequently, in this subsection that if uranium is not available from international suppliers, then even India's power reactors expanded by those currently under construction could meet little more in the way of electricity needs. Under such circumstances, it is possible that India could elect to terminate the first stage of its nuclear program, and convert its existing fleet of PHWRs to alternative uses, such as production of weapons-grade plutonium. The latter option is explored in Section 5.1.3.

Suppose the scenario in which 120 or fewer tonnes of UO_2 are produced every year, as consonant with the working estimate (Table 4.1) since 1998, is continued beyond 2008. This level of uranium production can, for example, meet the demand of the DHRUVA reactor (28 tonnes of uranium annually, at 75% capacity factor) and eight 220 MWe PHWRs operating at 55% capacity factor. To achieve a burn-up of 7000 MWd/tU at 55% capacity factor the 220 MWe PHWR needs to be refueled 165 times (8 bundle refueling on each occasion) in a year. This leads to a usage of 11.4 tonnes per 220 MWe PHWR annually (cf. Table 5.3 below). In terms of electrical production, this would be a very ineffective use of the current fleet of 18 operating power reactors and four under construction (Table 3.1).

At this level of operation, DHRUVA would produce 23.16 kilograms of weapons-grade plutonium annually, as computed with the burn-up code ORIGEN2.2. The PHWRs, if operated at anything less than 50% capacity factor, but with normal refueling sequence (300 days per year, at equilibrium; cf. Table 5.3), would produce plutonium of grade 20 % or lower (Pu-240/Pu-239), which is very high quality reactor-grade plutonium; material in this range of quality is sometimes

---

[1] This performance is not helped by the fact that transmission and distribution losses in India are the world's highest, with much of the loss due to theft; cf. Cleantech India, "India's Electricity Transmission and Distribution Losses," July 16, 2008, URL http://cleantechindia.wordpress.com/2008/07/16/indias-electricity-transmission-and-distribution-losses/, accessed June 27, 2009.

termed as "fuel grade." It is widely speculated that India tested a device composed of material of lower than weapons grade during the Pokhran-II tests.

Under the scenario of NO U.S.-India nuclear accord, some such state of affairs might have continued until uranium ores completely dried out. Alternately, a turn-around of the nuclear power industry could have come with discovery of new ore sites or availability of unsafeguarded natural uranium from suppliers outside the cartel known as the Nuclear Suppliers Group. This situation, or variants thereof, could have continued until the end of 2012, when India's uranium ore reserves were supposedly expected to get a boost from the present levels by newer (or revitalized) mining sites.[2] This scenario is further developed later in the present subsection.

Nonetheless, eventual civil nuclear power for India could come from its fast reactor program. India currently expects, by the end of 2012, to have an operational fast breeder reactor with fissile material being generated at a faster rate than ever previously produced. Eight PHWRs, at the low capacity factor envisioned above, will not significantly contribute toward satisfying electricity needs, but will give a significant further boost to the unprecedented production and accumulation of fissile material already envisioned above. This material conceivably could be used either for generation of civil energy, in fast breeders per stage 2 of the Bhabha program, or for weapons. These alternatives are, respectively, considered in the following two subsections.

Two new factors that could ease the pressure on uranium demand, and help in meeting electricity needs through nuclear power, are a rise in production by older mines and discovery of newer mines. From the available open source information, it is evident that a relentless effort to start newer mines is in progress. One of projects under pursuance is Turamdih site. The reserves of this mine have been assessed to a depth of 170 m; this mine was discovered in 1988 but was soon abandoned because of low ore-grade (0.02%). Work on it began later on 26th January 2004 and it is expected to be in service by January of 2009. For the first three years the ore production is planned to be 550 tonnes per day and later increased to 750 tonnes per day. A second site called Banduhurang mine where commissioning work was started on the same day (26th of Jan 2004) has been projected to start supplying domestic uranium by January of 2009.

There are no mining details available, so financial allocations were reverse calculated to estimate the rate of production shown in Table 5.1. Government of India has a budgetary commitment of $40.65 million (Rs.2000 crores) for Uranium Corporation of India Limited. These funds can bear the cost of 11.68 tonnes of UO_2 production. Data for compounds of uranium at various stages were calculated and are stated in Table 5.1.

From the calculated values of production and reserve the use of uranium under two different scenarios, 50% and 60% capacity factors, are analyzed in Table 5.2. As per the envisioned scenarios if 8 out of the 17 PHWRs are operated at 50% capacity factor, they can stretch their operating period beyond 2010. At any significantly higher capacity factor, India will run out of fuel even for these 8 unsafeguarded reactors. This suggests the decision to keep eight PHWRs designated

---

[2] A.B. Awati and R.B. Grover, Demand and Availability of Uranium Resources in India, Department of Atomic Energy, Mumbai, India (2005).

Table 5.1: Calculated Values of U_3O_8, UF_6 and UO_2 from Turamdih Mine.

Time Period	Total Ore (tonnes/day)	U_3O_8 (tonnes/yr)	U_3O_8 (tonnes/yr)[a]	UF_6 (tonnes/yr)[b]	UO_2 (tonnes/yr)[c]
2009–11	550	36.14	30.71	30.82	23.45
2012-...	750	49.28	41.88	42.02	31.98

[a] After 15% processing losses.
[b] After 20% conversion losses.
[c] After 0.8% fabrication losses.

as "strategic," under the U.S.-India accord, is a calculated gamble wherein these eight PHWRs are intended basically to provide fuel for the second-stage of Bhabha's three-stage-program. It is possible that some of these reactors could be returned higher capacity factor electrical production post-2011, especially if significant uranium production emerges from sites such as recently found in Andhra Pradesh[3] and Meghalaya.[4] As an alternative to the production from additional indigenous or non-NSG mines, India could have contemplated, in the event of failure of the delicate 2006-2008 negotiations with the U.S., the possibility to reengage in similar negotiations subsequently, for example around 2012.

Table 5.2: Fuel Production and Use in Tonnes for DHRUVA and Eight Unsafeguarded PHWRs with 260-day Refueling Sequence.

Year	Production	Use at 50% C.F.	Use at 60% C.F.
2009	143 (120+23.45)	130.5	Not possible
2010	143	130.5	to operate at
2011	143	130.5	this Capacity factor
2012	152 (120+31.98)	130.5	152

Under the scenario thus far envisioned in this subsection, only eight of India's currently operating fleet of 15 PHWRs (plus three additional under construction) would have been used to produce electricity and fissile material, until such time as additional uranium became available outside the sanction of the Nuclear suppliers Group. What would become of the remaining reactors during such an interim period? One possibility is that some of the (presumably older) plants could be shut down some time for refurbishment (e.g., coolant channel replacement), so as to be in prime condition at such time as India had access to additional uranium and only eight PHWRs are kept operational at 50% capacity factor. A second possibility would involve operation of more than eight

[3] Saraswati Kavula, "Uranium Mining and Nuclear Power," Jharkand Times, June 7, 2009, URL http://jharkhand.wordpress.com/2009/06/07/uranium-mining-in-nalgonda-and-tummalapalle-andhra-pradesh/, accessed June 27, 2009.
[4] Rahul Karmakar, "Hope for uranium mining in Meghalaya," Hindustan Times, May 12, 2007, URL http://www.hindustantimes.com/StoryPage/StoryPage.aspx?id=89d5ff03-4768-40b3-a89d-8bfba156ef96&&Headline=Hope+for+uranium+mining, accessed June 27, 2009.

reactors, at a corresponding lower refueling rate per reactor, so as to achieve the same burn-up and, therefore, production of electricity and quality and quantity of fissile material.

This second possibility has some technically based limitations. For one, an annual refueling sequence of 260 days, with 8 bundle shift each refueling day, is the base minimum required to keep the reactor critical and generate power, while providing adequate control of small reactivity transients. As a low excess positive reactivity of the core does not enable the control rods to be kept substantially (60-80%) inserted into the core there is a bare possible chance of achieving fast start-up following an unplanned shut-down. The loss of operating days (because of Xenon poisoning) due to shut-down may not be an issue because of the margin available to operate at higher capacity factor during days of power generation.

As the possibility of total uranium produced in 2009 and until 2012 is 120 tonnes from the existing sites and 23.45 tonnes from Turamdih mine, the feasibility of operating more than 8 reactors or higher capacity factors is eliminated. From 2012 onwards with higher budgetary allocations and production from Banduhurang mine, the capacity factor of the eight operating units possibly can be increased to 60%.[5] Thus, barring exceptional circumstances, without international fuel supplies India could neither operate more than 8 PHWRs nor reach economically prudent levels of power on the operating PHWRs.

5.1.2 FUTURE OF FAST BREEDER PROGRAM

The basis of the fast-breeder program is extraction of reactor-grade plutonium from the PHWR spent fuel. Under an average burn-up of 6750 MWd/tU for CANDU fuel bundles, 4 kilograms (precisely 4.108 kilograms) of reactor-grade plutonium (ratio of Pu-240/Pu-239 = 42%) can be extracted from one tonne of PHWR spent fuel. Given the historical operational reprocessing capability of 50 tonnes of spent fuel per year (before 1991 only one reprocessing plant of 100 tHM/yr existed and a capacity factor of 50% is assumed for it), India could extract 205 kilograms of plutonium annually from the spent fuel and, in total, could have produced 19.6 tonnes of separated plutonium from the beginning of its reprocessing capability. This quantity of plutonium is enough to support approximately[6] twenty fast breeder reactors of 500 MWe ratings if backed up by the required reprocessing capabilities. The fast-breeder program would not seem likely to have been deterred by failure of the cooperation agreement, as its fuel needs could have been (and can be) met from the vast reserves of the thermal reactor spent fuel, with sufficient reprocessing capability.

Whether the second-stage-power-program involving fast breeders is an economic success is at best a secondary consideration for the Indian scientific community because the fast breeder program can be justified on the basis of its ability to produce weapons-grade material. (See next subsection for more details.) Each FBR further enhances the plutonium and U-233 accumulation beyond the levels of present production rates.[7] This justification is not contingent upon immediate use of these

---

[5] Production at Banduhurang estimated by reverse calculations from known financial expenditures.

[6] More precisely, 19,600/850 ~23 reactors, based on an initial fuel load of 850 kilograms of plutonium. The latter was estimated via the MCNP transport code, and closely correlates to the Russian BN-600 reactor.

[7] A. J. Tellis, "Atoms for War," Carnegie Endowment, Carnegie Endowment for International Peace, Washington D.C. (2006).

materials to produce weapons. Rather, India's nuclear doctrine sanctions deferred retaliation as a deterrent to nuclear attack, so it is quite sufficient that these reactors are merely capable of providing material for weapons.

Given the estimated existing supply of plutonium, availability of fuel does not appear to be a limitation to ramp up the second-stage fast breeder program, but there are other obstacles. One is availability of sufficient financial capital, and another is lack of operational history for power plants based on fast reactors. As to the latter, fast reactors in France, Russia and Japan have experienced difficulties in handling of sodium. There is also a concern of debris clogging in the primary loop and core entry nozzles, due to the compact core size and restrictive dimensions.

The size of the Indian FBR program ultimately depends how far India determines to develop that program, notwithstanding the economic strain from it. India is a growing economy with extended nuclear infrastructure. A very aggressive goal for India would be startup of one plant per year under a closed fuel cycle with fast breeder reactors. Given FBRs with a doubling time of 20 years, this could be sustained on the basis of 20 breeders initially loaded with current supplies of plutonium, having thorium as fertile material, with subsequent (third-stage) reactors rather fueled by uranium-233. If the doubling time is less than 20 years, then the cycle can be sustained with fewer reactors initially fueled by plutonium.

The fundamental idea in the second-stage of the three-stage-program is to sustain the fast breeder program by recycling plutonium and producing U-233 that proceeds to fuel the third-stage reactors. Proliferation concerns are raised by the very fact that weapons-grade plutonium is produced and U-233, another weapons-usable material, is accumulated.

A non-proliferating fuel composition arguably would be one in which plutonium is produced at a higher ratio of Pu-240/Pu-239 and ratio of (U-233 + U-235)/U is less than 20%. A reactor system, which closely resembles these features, is thermal breeder reactor (TBR). In the past, TBRs such as the Shippingport Atomic Power Station operated for 5 years (1977-82) and fuel characteristics at the end (reported in 1987) showed a rise of 1.3% of fissile material content from that of initial loading.[8] TBR systems can become increasingly important under the present scenario of international collaboration and proliferation concerns. A study was carried out on a TBR design that can replace the fast breeders of the second-stage and integrate the second and third stages of the power program. Details of this TBR have been studied as "Alternate Reactor Systems for the Proposed Fuel Cycle" in Chapter 6.

India's commitment to the three-stage program is evident from the cumulative fuel cycle analysis in a flowsheet form until 2006 (Figure 4.5). There was a gross disregard to the reality of domestic uranium reserves in the process of rigid addition of power plants, even though a quicker advent to breeder program with indigenous capabilities of reprocessing and design of fast core in a closed fuel cycle might have, avoided the necessity of international cooperation in order to assure even minimal supply of electricity from nuclear energy.

---

[8] Rod Adams, "Light Water Breeder Reactor: Adapting a Proven System," Atomic Insights, Vol. 1, Issue 7, October 1995, http://www.atomicinsights.com/oct95/LWBR_oct95.html, accessed July 7, 2009.

5.1.3   EFFECT ON WEAPONS PROGRAM

According to the assessment of Figure 4.5, India would have had, at the end of 2006, sufficient plutonium to produce 69 implosion weapons (considering 6 kilograms of plutonium used for each weapon). If adequate uranium production is available to fuel DHRUVA, an addition of four weapons a year can be maintained. PHWRs operating at a lower capacity factor than the 50% contemplated in Section 5.1.1 above, but at the standard 306-day refueling sequence, could produce weapons-grade plutonium even if not operated deliberately for that cause. Operation at lower capacity factors does introduce certain technical challenges, but these seem solvable, as already discussed in Section 5.1.1.

Continued operation of the enrichment plant, as previously discussed, also assures the possibility of thermonuclear weapons and nuclear submarine core fuel.

If the international collaboration efforts had failed (or fail in the future), then most existing nuclear power plants could be closed down from power mode and plutonium production could be continued at the present level from the production reactors. The plutonium production rate from the PHWRs may, as already discussed, rise because of power reactors operating at low capacity. The ability to produce plutonium by maintaining the same on-power refueling sequence but operating it at a low capacity factor makes the PHWRs strategically important to India.

Contribution from breeders could be an additional boost to the weapons program. Each fast core can produce approximately 140 kilograms of U-233 every year.[9] Thus, under the base scenario of the preceding subsection (20 second-stage reactors, with a doubling time of 20 years), an amount of material more than equal to the total stockpile of weapons-usable material that has been added over four decades of struggle could be added every year. Under this scenario, the doors could conceivably remain open to cooperation, but it is difficult to conceive that India would agree to any terms that would roll back the clock to a lower rate of weapons-grade fissile material production.

Under a scenario of eight PHWRs operating outside safeguards at 50% capacity factor, the uranium fuel requirements for the next three years (2009-2011) would be 130.5 tonnes annually (Table 5.2). This is definitely possible within the present uranium reserves and an extrapolation of even the working estimate for uranium production. But this would entail production of 736 kilograms (92 kilograms Pu per reactor per year) of plutonium, for three years at a Pu-240/Pu-239 ratio of 20.3%, which is relatively high-quality reactor-grade material. As previously mentioned, India is presumed to have tested nuclear device with reactor-grade plutonium in Pokhran-II tests in 1998.[10]

5.2   PROJECTIONS UNDER ADOPTED U.S.–INDIA NUCLEAR ACCORD

The signing of the U.S.–India nuclear agreement brings fourteen of the twenty-two Indian reactors operating or under construction under India-specific safeguards. These safeguards are somewhat

---

[9] Government of India, Department of Atomic Energy, Annual Report 2001-2002, Mumbai, India (2002).

[10] The Nuclear Weapon Archive, "India's Nuclear Power Program," http://nuclearweaponarchive.org/India/IndiaOrigin.html, (March 2001).

different from those applied to Nuclear Weapon States (NWSs) and Non Nuclear Weapons States (NNWSs), as defined under the NPT. Four reactors already were under IAEA safeguards, or scheduled to be. India agreed that all of its future *civil* reactors would be under safeguards, but it also reserved the sole right to designate which of its future reactors would be "civil." The most controversial research reactor of the non-proliferation debate, CIRUS, is to be shut down by 2010. CIRUS produced approximately 1/3rd of the total weapons-grade plutonium that India possesses as of now. (That is, 187.2 kilograms out of the total of 633.5 kilograms produced between 1964 and 2006; cf. Section 4.6). The implications of this cooperation can better be understood by analyzing its projections on the nuclear power program, fast breeder program and weapons initiative.

5.2.1   FUTURE OF NUCLEAR POWER PROGRAM

Under the cooperation agreement India would have eight unsafeguarded PHWRs, once all reactors currently under construction are completed. The uranium fuel rods used in these heavy-water cooled and moderated nuclear power plants can be reprocessed to extract plutonium for the fast breeders of the second stage, as already discussed in the preceding section. While operating in power mode the uranium fuel remains in the reactor for 300 to 310 days to reach a burn-up of 7000 MWd/tU resulting in 2600 grams of Pu-239, 1108 grams or more of Pu-240, 306 grams or more of P-241, and about 90 grams of Pu-242 for every tonne of natural uranium fuel. Plutonium extracted from spent fuel irradiated to such burn-ups is not desirable for use in nuclear weapons because of a high concentration of Pu-240. Table 5.3 details the fuel requirements and characteristics of the resulting spent fuel for a 220 MWe PHWR operated at varying capacity factors with a refueling sequence of 300 days per year (eight fuel bundles replaced per refueling day). (Note 1: The last two columns refer to operation at the indicated capacity factor, but with refueling sequence adjusted to provide a burn-up of 7000 MWd/tU). (Note 2: If the 220 MWe PHWR is operated with 300 days refueling in a year to produce Pu-240/Pu-239 ratio of 6% then the burn-up achieved is 1066.67 MWd/tU with total Pu production of 908 grams). These data are used in the following two subsections to assess India's options for use of the plutonium in the spent fuel derived from operation of its strategic PHWRs.

Under the agreement, India will have eight PHWRs, six of 220-MWe capacity and two of 540-MWe capacity, available for consumption of its indigenous uranium, with no restrictions on the usage of the materials in the resulting spent fuel. This gives a slightly higher total capacity than the eight 220-MWe PHWR scenario considered in Section 5.1.1, but the net options for consumption (at slightly lower capacity factor) of all of the indigenous uranium and the results in terms of production of power and spent fuel are identical.

The range of options available to India, either with or without the cooperation, can be conveniently displayed graphically via a "latency diagram" such as shown in Figure 5.1. Here the "uranium constraint" derives from the assumption that 250 tonnes of indigenous uranium will be consumed annually at a burn-up of either 6750 or 1067 MWth-days per tonne U, and that the higher (lower) of these burn-ups will produce 4.108 kilograms of reactor-grade plutonium (respectively 0.908 kilo-

Table 5.3: Plutonium Production and Uranium Use from Operation of a 220 MWe PHWR at Various Capacity Factors, for a Refueling sequence of 300 Days per Year and Eight Fuel Bundles Extracted per Refueling Day.

C.F. (%)	Burn-up MWd/tU	Pu-239 (gms/tU)	Pu-240 (gms/tU)	Pu-241 (gms/tU)	Ratio of Pu-240/Pu-239 (%)	Pu (gms/tU)	Total Pu produced (kg/yr)	Nat. U used (tonnes/year) at 7000 MWd/tU burn-up	Refueling days per year at 7000 MWd/tU burn-up
1	70	68.92	0.280		0.4		3	0.004	3
5	350	324.6	6.536		2.0		12	0.094	15
10	700	605.4	24.16		4.0		24	0.377	30
11	770	657.1	28.8		4.4		26	0.456	33
12	840	707.5	33.78		4.8		28	0.543	36
13	910	756.6	39.08		5.2		30	0.637	39
14	980	804.5	44.69		5.6		32	0.739	42
15	1050	851.1	50.61		5.9		34	0.848	45
20	1400	1068	84.35	5.674	7.9	1158	44	1.5	60
25	1750	1262	124.4	10.24	9.9	1397	53	2.4	75
30	2100	1437	170.1	16.51	11.8	1624	61	3.4	90
35	2450	1594	221	24.66	13.9	1840	69	4.6	105
40	2800	1738	276.9	34.88	15.9	2050	77	6.0	120
45	3150	1869	337.7	47.36	18.1	2254	85	7.6	135
50	3500	1988	403.3	62.31	20.3	2454	92	9.4	150
55	3850	2096	473.4	79.9	22.6	2649	100	11.4	165
60	4200	2194	547.8	100.3	25.0	2842	107	13.6	180
65	4550	2282	626.3	123.7	27.4	3032	114	15.9	195
70	4900	2360	708.4	150	30.0	3218	121	18.5	210
75	5250	2429	793.5	179.3	32.7	3402	128	21.2	225
80	5600	2490	880.8	211.6	35.4	3582	135	24.1	240
85	5950	2541	969.6	246.4	38.2	3757	142	27.2	255
90	6300	2584	1059	283.6	41.0	3927	148	30.5	270
95	6650	2620	1147	322.6	43.8	4090	154	34.0	285
100	7000	2649	1235	362.9	46.6	4247	160	37.7	300

grams of weapons-grade plutonium) per tonne of uranium. These plutonium productions are from calculations (results not shown) similar to those underlying the fourth column from the right in Table 5.3. Similar data underlie the capacity curves, except now the defining factors are 220 MWe capacity with (without) the agreement, assumed thermal efficiency of 0.3 and capacity factor of 100%.

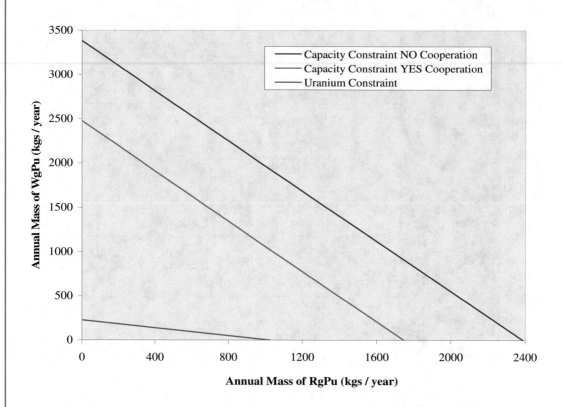

Figure 5.1: Latency capability with and without cooperation, and with uranium constraint of 250 tonnes annual indigenous production.

In principle, India can choose to produce weapons-grade and reactor-grade plutonium (of the specified qualities) at any point that is below and to the left of both the uranium constraint and the operative capacity constraint. For the 250 tonnes annual indigenous production of uranium assumed in Figure 5.1, the relevant constraint is that associated with uranium production, either with or without the cooperation, as would be true at any lower level of uranium production (e.g., the working estimate considered above). The 250 tonnes of natural uranium could lead to production of 1027 kilograms of reactor-grade plutonium annually (x-intercept of the uranium constraint in Figure 5.1), should India choose to give priority to production of electricity and therefore maximize production of reactor-grade plutonium.

The difference between the two y-intercepts of the two capacity constraints in Figure 5.1 is a graphic measure of how much India's capability to produce weapons-grade plutonium would be reduced by the cooperation, if availability of uranium were not an issue. On the other hand, if the uranium supply were increased enough so that the uranium constraint (always parallel to its position in Figure 5.1) intercepts the capacity constraint under the cooperation at a positive ordinate, then the value of that ordinate would represent the amount of weapons-grade plutonium India could produce with no loss in electrical production.[11]

5.2.2   FUTURE OF FAST BREEDER PROGRAM

The nuclear treaty between U.S. and India mandates that all current and future nuclear power plants that India declares as civil will be subject to safeguards in perpetuity, and will be eligible to be fuelled from the international market. India's initial declaration of civil facilities, as pronounced in its "separation plan," designates as civil 14 of the 22 nuclear power plants that are currently either existing or under construction. The plutonium fuel for India's second-stage fast breeder reactor must come from plutonium previously produced, most in PHWRs, or plutonium to be produced in the future, the latter either from indigenous or international uranium.

As regards either existing plutonium or future plutonium from indigenous uranium, the agreement seems to bring changes that are, at most, very marginal relative to the situation that would prevail without the agreement, as already discussed in Section 5.1.2, and further in conjunction with Figure 5.1. Specifically, the amount and nature of the existing plutonium and indigenous uranium supply are unchanged by this agreement. The reprocessing facilities available for spent fuel from indigenous uranium likewise is unchanged by the agreement, as India has agreed to build a new reprocessing facility for any spent fuel derived from internationally supplied uranium.

The most significant change, as regards indigenous materials, seems to be that if India wishes, as seems likely, to maintain the option of using materials derived from its indigenous uranium for weapons purposes, then it will have fewer reactors to process those materials than would have been the case without the agreement. However, as argued in considerable detail above, the eight reactors retained as "strategic" seem more than adequate, for any conceivable practical purpose, to process the amounts of indigenous uranium India is likely to have over the next five years or so. Access to additional uranium supplies not subject to safeguards, indigenous or otherwise, could cause a shortage of reactors to process those materials. That problem could eventually be solved by constructing additional nuclear power plants using indigenous technology. On the other hand, access to the international uranium market could reduce the incentive to seek uranium that would be free from safeguards requirement. These are questions of intent rather than capability, so are not subject to analysis from a strictly technical viewpoint. Some similar considerations, but more directly related to potential impact on the weapons program, are further discussed in the following subsection.

[11] See Paul Nelson, Taraknath V.K. Woddi and William S. Charlton , "A Framework of Analysis of the weapons Implications of the U.S.-India Nuclear Accord," NSSPI Report No. 07-010, Nuclear Security Science and Policy Institute, Texas A&M University, October 10, 2007, http://nsspi.tamu.edu/projects/p0/p0_pub11.pdf, accessed July 4, 2009, for further instances of such "latency diagrams."

The most interesting consequences of the agreement for India's fast breeder program revolve about the fact that spent fuel derived from internationally supplied uranium now will, in principle, be available as a source of plutonium fuel for the fast breeders. Such plutonium, and therefore presumably the fast reactors it fuels, will be subject to safeguards. Safeguarding the material presumably will not be a significant issue, as India has long accepted safeguarding of materials derived from imported material, under the aegis of the "principle of pursuit." Safeguards on the reactors themselves might be more sensitive, as doubtless there will be design features of those reactors that India regards as commercially valuable property. The higher the quality of the spent fuel (i.e., lower the Pu-240/Pu-239 ratio) that India deems to be required for its FBRs, the more sensitive are likely to be any related negotiations with the U.S. and the international community, because higher quality also is more desirable for weapons use. Such issues are likely to engage Indian and U.S. diplomats, and their advisory nuclear technologists, for some time into the future.

5.2.3   FUTURE OF WEAPONS PROGRAM

The cooperation poses more-or-less the same somewhat minimal issues for India's weapon program as for its fast reactor program, as already discussed in the preceding section. The effects that have a technical basis are as follows.

Suppose first that one presumes India's intent is to give first priority for its indigenous uranium to production of electricity. In such a case, if India's indigenous production of uranium should become slightly larger than required to fuel the eight strategically reserved reactors at the high burn-up optimal for electricity production, then the "excess" could be employed at a low burn-up as most desirable to produce plutonium most desirable for weapons use. In Figure 5.1, this would be represented graphically by intersection of the uranium constraint with the capacity constraint under the cooperation at a positive value of the ordinate. Some see this as potentially "freeing up" India's indigenous uranium for weapons use, in the amount corresponding to that ordinate. Note that this would occur only under circumstances of availability of indigenous uranium, or other supplies not subject to international safeguards, in considerable excess of the "working estimate" most of the preceding discussion has been based on.

Suppose now, on the other hand, that India elects to give priority to production of weapons-grade plutonium from its strategically reserved PHWRs. At some hypothetical even higher future production of indigenous Indian uranium, the availability of reactors to process that uranium would be the constraining factor in India's production of uranium. The potential for that to become a constraining factor is increased by the cooperation, to the extent measured by the differences of the y-intercepts of the two capacity constraints in Figure 5.1. Of course, this eventually could be overcome by construction of additional strategically reserved reactors, just as discussed in the preceding subsection for the fast breeder program.[12]

---

[12] See Nelson, Woddi and Charlton, op cit., for a more detailed analysis of the full range of weapons vs. electricity options available to India. This work attempts to incorporate further subtleties, such as limitations associated to rate of refueling, but others remain inaccessible (e.g., possible production of plutonium at some intermediate range of qualities).

Additionally, impact on weapons program could affect the commitment of India to a search for uranium not subject to international safeguards, just as discussed in the preceding subsection in the context of the fast breeder program. At the same time, denial of the agreement could have affected the extent to which India committed its scarce uranium resources to defense as opposed to electricity. In a worst-case scenario, such denial could have even (eventually) led to abandonment of a civil nuclear program deemed to be doomed by lack of international cooperation, in favor of focusing all of India's extensive nuclear capabilities exclusively on defense.

The ultimately important fact is that all of these possibilities are, at best, remotely hypothetical. Under by far the most likely scenarios, the agreement essentially is a non-event, with respect to India's nuclear weapons program. One suspects this is by very careful design, backed by very close negotiation. Had the agreement significantly assisted that program, it could not have been seriously contemplated by the international community, as led by the U.S. Had it the potential for significant adverse impact on that program, then no Indian government could have supported it.

CHAPTER 6

Alternate Reactor Systems for the Proposed Nuclear Fuel Cycle

An alternate reactor system that would further the cause of separation of strategic and civilian nuclear fuel cycle and enable efficient use of resources under safeguards and strengthen the non-proliferation cause has been developed. The alternate reactor system was optimized for efficient utilization of domestic thorium and international uranium resources. The proposed alternate reactor system is a breeder with thermal spectrum and CANDU-type core geometry. This Thermal Breeder Reactor (TBR) was designed and optimized to fit the proposed fuel cycle of India. The suitability of the alternate reactor system in the power program was synthesized on specific metrics that effectively represent the objectives. This system was also studied in comparison to the presently planned PH-WRs. The objective of developing a TBR was to convert Th-232 to U-233 at an optimized breeding ratio and simultaneously achieve high burn-up of the fuel. The end-of-cycle (EOC) fuel composition was ensured to be proliferation resistant by optimizing the system with the specified metrics. The metrics to measure the objectives are:[1]

- Mass (kilograms) of uranium used per unit (MWday) of electricity production,

- Pu-240/Pu-239 ratio to establish the reactor-grade nature of plutonium,

- Total plutonium production and

- Presence of less than 20% fissile uranium (U-233 + U-235) in the total uranium mass.

6.1 THERMAL BREEDER REACTOR DESIGN CONCEPT

Thermal breeder reactors primarily operate on the basis of neutron absorption by fertile isotopes in a thermal spectrum, producing more fissile fuel than they consume. Earlier studies on breeders have shown that the absorption cross-section is an important factor in choosing fertile material for the core.[2] The fact that Th-232 breeds U-233 through neutron absorption and successive beta decays with higher neutron absorption cross-section than U-238 was an overriding factor favoring thorium in thermal breeders.[3] However, such breeders raise the concern of proliferation as U-233 is weapon usable material.

---

[1] W.S. Charlton, Ryan F. Lebouf et al., "Proliferation Resistant Assessment Methodology for Nuclear Fuel Cycles," Nuclear Technology, American Nuclear Society, Vol .157 (Feb 2007).

[2] A.M. Perry and A.M. Weinberg, "Thermal Breeder Reactors," Annual Reviews Nuclear Science, p. 317 (1972).

[3] B.I. SPINRAD, Alternative Breeder Reactor Technologies, Annual Reviews Energy, p. 147, Department of Nuclear Engineering, Oregon State University, Corvallis, (1978).

If U-233 and U-235, both weapon usable materials, are produced or present in a mix with various transuranic and transplutonic isotopes and are less than 20% of the total uranium content then it would be hard to separate the weapons-grade material from the spent fuel. A high-cost reprocessing facility followed by an enrichment process would be required to separate the fissile content. The intermixing in the spent fuel is obtained with a homogeneous mixture of thorium and low enriched uranium oxide as fuel in the fresh core. This limits the percentage of thorium that can be used in the fuel elements to below 94% with the balance being natural or low enriched uranium. The incorporation of some low enriched uranium provides another safety advantage relative to all-thorium fuel because the lower absorption cross-section for epithermal neutrons in pure Th-232, reduces the negative power co-efficient incase of a power transient. But too much uranium in the fuel will result in a higher concentration of plutonium produced by the fertile isotope U-238. In a thermal breeder reactor of the suggested design, the low enriched fuel reaches high burn-up and achieves higher Pu-240/Pu-239 ratio for the EOC fuel. The blanket with half the size of the driver also breeds only reactor-grade plutonium, in lower quantity. Also stated in the previous section, the increased ratio of Pu-240/Pu-239 reduces the proliferation risk because the spontaneous fission neutron yield of Pu-240 can prematurely detonate an assembled weapon.

6.2   THE THEORY OF BREEDER REACTORS

A fissile production rate that exceeds the fissile consumption rate can be attained through an appropriate combination of fissile and fertile mass arranged in a suitable geometry, together with proper reprocessing schedules.[4]. The ratio of the mean rate of fissile material produced to the mean rate of fissile material consumed is defined as the "Breeding Ratio" (BR). A BR > 1 is a self sustaining reactor system and is called a breeder. If BR < 1 then the reactor is Converter Reactor and the Breeding Ratio is usually referred to as "Conversion Ratio" (CR). BR can be shown to be equal to the ratio of the fissile material produced (FP) to fissile material destroyed (FD) during a fuel cycle (i.e., between periodic refueling or at the end of average burn-up): [5]

$$BR = \frac{FP}{FD} \tag{8}$$

The breeding gain is given by

$$G = BR - 1. \tag{9}$$

Substituting equation (8) into equation (9) yields:

$$G = \frac{FP}{FD} - 1 \ \mathrm{Or} \ G = \frac{FP - FD}{FD} \tag{10}$$

[4] B.I. SPINRAD, Alternative Breeder Reactor Technologies, Annual Reviews Energy, p. 147, Department of Nuclear Engineering, Oregon State University, Corvallis, (1978).
[5] A.M. Perry and A.M. Weinberg, "Thermal Breeder Reactors," Annual Reviews Nuclear Science, p. 317 (1972).

In the numerator of equation (10), the fissile material produced minus the fissile material destroyed is basically the total fissile content at the end-of-cycle (EOC) after subtracting the initial loading. Thus, it can also be inferred as ratio of the difference of fissile materials at the end and beginning of cycle (BOC) to the fissile material destroyed. The breeding gain

$$G = \frac{FEOC - FBOC}{FD} \tag{11}$$

is normally accounted for at the end of the one year but since the suggested breeder design is an on-power refueling reactor, it is also computed at average burn-up.

Like any other reactor the physics of thermal breeding reactors are driven by neutron economy. A nuclear reactor can breed over a broad energy spectrum but adequate breeding ratios can only be realized in a certain energy range. A high breeding gain is attained with a fast neutron spectrum, but a low breeding gain on a faster fueling cycle characteristic of some thermal breeder designs like that of on-power refueling CANDU-type cores is a superior design as shown through this study.

The terms universally known in the fission process are 'v', the number of neutrons produced per fission, 'η', the number of neutrons produced per absorption and 'α', the capture to fission ratio (σ_c/σ_f).

These parameters are related by,

$$\eta = \frac{v\sigma_f}{\sigma_f + \sigma_c} = \frac{v}{1 + \sigma_c/\sigma_f} = \frac{v}{1 + \alpha}, \tag{12}$$

The parameters v and α are measured quantities, while η is a derived quantity. In a thermal neutron energy spectrum v is fairly constant up to the energy range of 1 MeV for each of the primary fissile isotopes, while α varies considerably with energy and between isotopes. This behavior of v and α leads to variations in η over energy range. In a thermal reactor, the energy of the neutron increases with rise in moderator temperature. This results in capture of neutrons in resonance cross-sections leading to reduced η. Increases in fuel to moderator ratio also enhances the importance of the near thermal and epithermal neutrons, thus reducing the average value of η. Consider a simple neutron energy balance where one neutron is absorbed by a fissile nucleus in order to continue the chain reaction and 'L' neutrons are lost unproductively by parasitic absorption (capture in structures, coolant, control rods, poisons and fission products) and also by leakage from the reactor. The number of neutrons left for capture by fertile nucleus is $[\eta - (1 + L)]$. So to produce a fissile nucleus from a fertile nucleus after a fission event we need to have $[\eta - (1 + L)] \geq 1$ or $\eta \geq 2 + L$. The quantity $[\eta - (1 + L)]$ is basically fissile nuclei produced to fissile nuclei destroyed, thus making the breeding ratio (BR) equal to $[\eta - (1 + L)]$. Taking into account the neutrons lost to compensate reactivity for moderator and coolant expansion the breeding ratio actually achieved is

$$BR = [\eta\varepsilon - (1 + L)], \tag{13}$$

where ε is the correction factor for the moderator and coolant temperatures.

From equation (13) it can be concluded that maximum BR possible is $\eta\varepsilon - 1$ or $\bar{\eta} - 1$, where $\bar{\eta} = \eta\varepsilon$.

Thus, the maximum breeding ratio

$$BR_{\max} = \bar{\eta} - 1 \tag{14}$$

represents the quantity of fissile material produced in a certain time or at average burn-up. Now to calculate the actual quantity of fissile material produced, a gain factor needs to be formulated.

A reactor gain factor (RGF) is a representative number for a particular breeder reactor producing fissile material in excess of its own fissile inventory to fuel an identical reactor at the end of its average burn-up cycle. Given the average burn-up of the fuel type, the required doubling time varies accordingly. In this study, RGF has been defined in terms of initial fissile inventory (M_0 kilograms) used by the reactor and the fissile material (M_g kilograms) gained at the end of the average burn-up.

$$RGF = \frac{M_g}{M_0} \tag{15}$$

Where, M_g kilograms is the produced fissile material content (^{233}U, ^{235}U, ^{238}N$_p$ and Pu)

$$M_g = FEOC - FBOC \tag{16}$$

From equations (11) and (16) we have

$$M_g = FD * G \tag{17}$$

Or, $M_g = G * (1 - \alpha) * \quad$ (Fissile mass fissioned until the average burn-up) $\tag{18}$

The fissile mass fissioned until the average burn-up is computed as power in watts of the reactor core multiplied to 2.93 X 10^{10} fissions / watt-seconds and time to reach average burn-up stated in seconds, times the molar mass of 238 gms / gm-mole. The correction factor for actual operation is considered by multiplication of the reactor operational capacity factor. The value obtained is divided by 6.023 X 10^{24} atoms / gm-mole to get the mass in grams.[6]

The other measure to judge the fissile material production is the Reactor Doubling Time (RDT), the time required by the particular breeder reactor to produce fissile material in excess of its own fissile inventory. Hence, it is the time necessary to double the initial load of fissile material to fuel an identical reactor, defining M_{gd} kilograms as the time averaged difference between the fissile inventory at the beginning of the year and the fissile inventory at the end of the year:

$$RDT = \frac{M_{gd}}{M_0} \quad \text{(in years)} \tag{19}$$

and $M_{gd} = FEOC - FBOC,$ computed at the end of one year $\tag{20}$

[6] S. MCLAIN and J.H. MARTENS, Reactor Handbook, Interscience Publishers, New York (1964).

From equations (11) and (20) we have

$$M_{gd} = FD * G \tag{21}$$

$$\text{Or, } M_{gd} = G * (1 - \alpha) * \text{ (Fissile mass destroyed in a year)} \tag{22}$$

The fissile mass fissioned is calculated in a similar manner as above with time being considered as one year instead of time to reach average burn-up.

6.3   REACTOR PHYSICS SIMULATIONS

6.3.1   THE SCALE CODE SYSTEM

The whole core 3D model of TBR, in CANDU-type core geometry with various combinations of fuel compositions, was created with SCALE version 5.1 modular code systems. This SCALE modular code system is developed and maintained by Oak Ridge National Laboratory (ORNL) and is widely accepted around the world for criticality safety analysis and depletion calculations. Figure 6.1 shows the modules in a sequential pattern for executing the desired input.

The SCALE control module TRITON was used to perform depletion calculations. TRITON couples KENO V.a 3D Monte Carlo transport code with the well known SCALE point depletion and decay module ORIGEN-S, which tracks more than 1500 nuclides. All TRITON models used the 238-group ENDF/B-VI cross section library and the BONAMI, WORKER/CENTRM/PMC modules for cross section processing. The BONAMI module provides resonance corrected cross sections in the unresolved resonance range, and WORKER/CENTRM/PMC modules provide resonance corrected cross sections in the resolved resonance range. CENTRM is a one-dimensional discrete ordinates code that computes space-dependent, continuous-energy neutron spectra. PMC uses the spectra to collapse the continuous-energy cross section data to multi-group (238 group) data for use by KENO V.a.

The KENO V.a criticality transport code determines the effective multiplication factor. The KENO post-processing utility KMART is used to extract fluxes, determine power distributions, and collapse cross sections to the three-group form required by COUPLE and ORIGEN-S for depletion calculations. The OPUS module provides the ability to extract specific data from ORIGEN output libraries, perform unit conversions, and generate plot data for post-calculation analysis[7]. Illustrated in Figure 6.1 is the calculation flow path during TRITON depletion calculations. The KENO V.a models were run with 4050 neutron generations, skipping the first 50 generations. There were 5000 neutrons in each generation, resulting in a total of 2 million neutron histories.

6.3.2   SIMULATION DETAILS AND DESIGN ITERATIONS

A series of stepwise iterations with different fuel configurations for driver and blanket fuel suitable to CANDU-type geometry were done to optimize TBR. The optimized breeder core TBR-1 comprises

---

[7] OAK RIDGE NATIONAL LABORATORY, "SCALE: A Modular Code System for Performing Standardized Computer Analyses for Licensing Evaluation," ORNL/TM-2005/39, Oak Ridge National Laboratory, Oak Ridge, TN (2005).

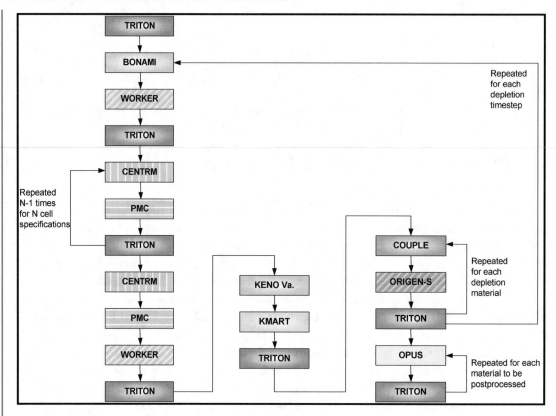

Figure 6.1: Execution path for the TRITON analytical depletion sequence.

of 80:20 uranium-thorium in driver fuel and 80:20 proportion of spent fuel from PHWRs and thorium in blanket assemblies. The driver fuel portion of the core (304 out of 456 channels) contains 3% of U-235 by mass. The radial blanket assemblies surrounding the driver fuel are made of 80 % spent fuel of CANDU-6 core after the fission products were removed and the remaining 20% is thorium. Table 6.1 and 6.2 details the fuel composition in the core. This combination of fuel achieves nearly 90% of the burn-up of a CANDU-6 reactor with slightly higher rate of depletion from fresh core to the end-of-life (shown in Figure 6.2) in the first cycle. The average burn-up achieved for the final two breeder core designs TBR-1, 2 and CANDU-6 fuels were 18.72, 19.68 and 22.08, GWd/tHM respectively. The TBR-2 core KCODE and depletion modules are also reported for comparison purposes. TBR-2 driver fuel consists of pure uranium metal and blanket being of 100% thorium fuel. As evident in this study, spent fuel from the TBR-2 is not proliferation resistant and does not fulfill the necessity of burning the plutonium accumulated from thermal reactors. The

Table 6.1: Driver Fuel Configuration for the TBR-1 and 2		
	TBR - 1	TBR - 2
Materials	Weight %	Weight %
O-16	9.6	0
U-235	3	3
U-238	67.4	97
Th-232	20	0

blanket being made of 100% thorium also leads to accumulation of weapons-grade fissile material U-233 in a chemically separable form.

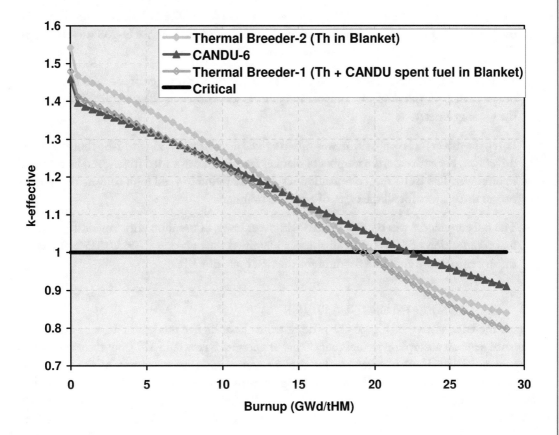

Figure 6.2: K- effective versus burn-up for thermal breeders and CANDU-6 cores.

The steps of simulation, assumptions and modeling criteria involved in achieving the optimized thermal breeder reactor fuel configuration is as stated:

1. The total quantity of the heavy metal was kept equal to that of the modeled CANDU-6 core because of similarity in material properties of the fuel, core size and structure.

2. To start with seven outer layers of the core were filled with blanket fuel assemblies consisting of 10% thorium and 90% spent fuel of PHWRs.

3. The enrichment of the driver fuel was increased from natural uranium to 20% U-235 in evenly distributed steps of 5%. The limit of 20% was set so as not to exceed the weapons-grade enriched uranium level.

4. The end of life (EOL) fuel composition was then compared against the initial fissile matter for all of the five cases.

5. In the next step, the thorium percentage was increased from 10% to 70% in steps of 10% for cases with 5% and higher enriched uranium. The minimum limit of 5% U-235 was chosen on the basis of achievable burn-up close to that of CANDU-6 core.

6. After comparing the EOC fuel composition for all the cases, the range of uranium enrichment of less than 10% and thorium composition of 20% was chosen as the most feasible point for the given geometry.

7. Later, the driver fuel composition was varied from 3% to 10% U-235 with 20% thorium in the driver fuel. For each case, the number of blanket assemblies was varied from seven layers down to one layer. The EOC fuel composition results were found to yield high conversion ratio for two to three layers for all the cases of enriched uranium.

8. The full core model was then simulated with finer levels of uranium enrichment. Finally, the fuel composition of 4.25% enrichment was identified thus making it 3% U-235 by mass for the driver fuel core and 20% thorium for the first cycle of TBR.

6.3.3   RESULTS FROM SIMULATIONS

The core model as shown in Figure 6.3 is designed to have higher moderator-to-fuel-atom ratio to enhance neutron absorption in fuel that occurs at energies below 0.45 eV. Low absorption cross section material like D_2O is necessary for coolant and as moderator because of the low fissile content of slightly enriched U-235 fuel. Limiting neutron loss due to leakage in thermal breeders is largely a matter of economics. Leakage has been reduced by surrounding the active core with blanket of appropriate thickness. The optimal thickness of the blanket (2-3 assemblies) was attained by balancing the cost of incorporating more assemblies against the benefits of additional neutrons saved.

The blanket assemblies are cooled and moderated with D_2O. The driver fuel and blanket assemblies are surrounded with D_2O reflector both radially and axially. Each driver fuel and radial blanket assembly (as shown in Figure 6.4) consists of a calandria tube (CT) surrounding a pressure

Table 6.2: Fuel for Radial Blanket Assemblies of TBR–1			
Materials	Weight %	Materials	Weight %
U-234	0.0000289	Np-237	0.0083050
U-235	0.2456285	Am-241	0.0006869
U-236	0.2085262	Am-243	0.0016323
U-238	79.077783	Cm-242	0.0002466
Pu-238	0.0019075	Cm-243	0.0000032
Pu-239	0.2470208	Cm-244	0.0002342
Pu-240	0.1533233	Cm-245	0.0000033
Pu-241	0.0368484	Am-242m	0.0000078
Pu-242	0.0178140	Np-237	0.0083050
U-234	0.0000289	Th-232	20

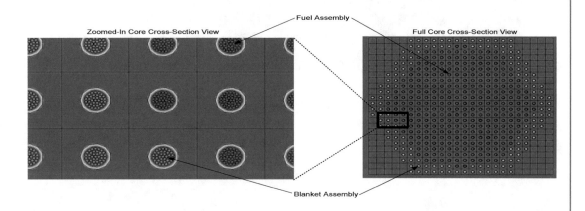

Figure 6.3: Cross-sectional view of the TBR model (KENO 3D).

tube (PT) with twelve axially placed cylindrical fuel bundles inside it. The annular space between the CT and PT is filled with CO_2 for tube failure detection. Each fuel bundle is made of 37 circular pincell, Zircolay-4 cladded fuel rods 50 cm long. The gap between fuel and clad is filled with nitrogen to diminish fuel clad interaction and provide lubrication to the fuel pellets.

The modeled thermal breeder has an average reactivity drop of 0.8 mk per day. The core is kept operational by online refueling. To maintain the excess reactivity it needs to be refueled everyday with 10 fresh bundles (shown as in Figure 6.5). This is subject to the condition that core heat is generated only from the driver fuel assemblies. A detailed refueling sequence for online fueling and blanket bundles replacement would depend on the specific heat distribution of the core. The heat distribution of the core for blanket and driver assemblies can be altered by incorporating heterogeneity in the core geometry. This change would need detailed analysis with respect to the blanket power, quench

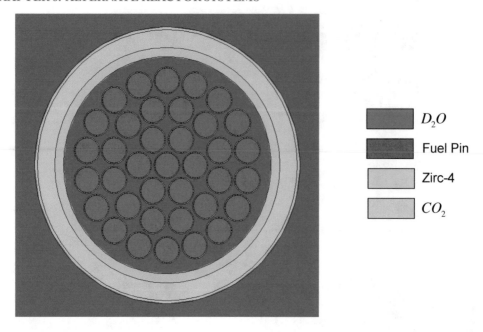

Figure 6.4: Top view of the thermal breeder fuel assembly model (KENO 3D).

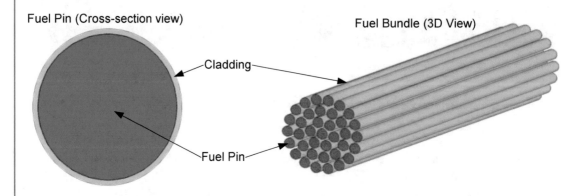

Figure 6.5: Image of the thermal breeder fuel pin and 37-pincell bundle (KENO 3D).

time and reflooding sequence during LOCA. A higher percentage of blanket power would also call for intricate reactivity regulation and protection system along with revised refueling sequence.

Continuous burn-up of the fuel results in production of fissile isotopes of the likes of Pu-239, Pu-241 and U-233, some of which fissions while some accumulates until the end of average burn-up. Material quantities for the driver and radial blanket fuel assemblies at the beginning and end of

Isotopes	Driver Fuel Configuration (kilograms)		Radial Fuel Configuration (kilograms)	
	BOC	EOC	BOC	EOC
U-235	2538	589	112	78
U-233	0	195	0	34
Np-237	0	10	3.77	4.38
Th-232	16922	16500	9083	9037
Pu-238	0	2	0.88	1.29
Pu-239	0	196	112	107
Pu-240	0	99	70	78
Pu-241	0	27	17	18
Pu-242	0	10	8	11
Total Pu	0	529	208	215
Fissile Content	2538	1118	320	327
Fissile Content Equivalent	2538	1238	332	358
Fissile Content Destroyed	–NA–	1948	–NA–	34

Table 6.3: Fissile Material Accounting for TBR–1

cycle of the first burn-up cycle are given in Table 6.3. The spent fuel of the driver and blanket are intermixed and U-235 was added to bring the fissile content in the driver fuel back to 3%. Because of the change in fuel isotopes, the average burn-up reached for the second cycle is different from the first cycle. Repeating the process of irradiation followed by making up for the loss in fissile content, it takes eleven cycles to reach a stage of breeding ratio that is greater than one. The equilibrium state achieved has a breeding ratio of 1.04 with EOC fuel composition rich in fissionable plutonium isotopes and U-233.

As the fuel for the blanket is made available from the accumulated spent fuel of PHWRs, the quantity of material can be judged as freely available to the cycle from its thermal reactor predecessors. This quantity of fissile material consumed by the blanket can also stay out of the calculation because it is has been already paid for. At the end of 585 days, the power produced by the blanket is 7.7% and rest by the driver fuel. Fissile content equivalent takes into account that Pu-241 worth is 1.5 times more than Pu-239 primarily because of higher fission cross-section.

Applying equation (8) for the data of the first cycle from Table 6.3, we have FP = 1596 kilograms (1238 + 358) and FD = 1948 kilograms.

Thus, the conversion ratio for the first cycle of TBR-1 is 0.819. To reach the average burn-up it takes 585 days of full-power-day operations and for fissile materials accumulation for one full core it needs to operate for 585/0.819 days or more in first stage. This is close to 715 days at 100% capacity factor. Taking a realistic annual capacity factor of 75%, the first cycle of thermal breeder would produce a full load of fuel in every 2.6 years. The fuel produced would be a mix of U-233, U-235 and plutonium. After the first cycle of core operation, the spent fuel is reprocessed and reloaded in the driver fuel assemblies along with required additional U-235 to bring the fissile content to 3% in the driver part of the core. The process of irradiation and reprocessing reaches an equilibrium state after eleven cycles of reactor operation spanning 11 years. The proportion of plutonium and uranium in the EOC fuel is 2.4% and 1.69%, respectively, in the equilibrium cycle. The doubling time for this reactor in a fuel cycle is dependent on the number of reactors and the rate of reprocessing of the spent fuel. The simulations and calculations above indicate that a series of eleven TBRs built at a rate of one per year can realize an equilibrium breeding fuel cycle in eleven years. The next step in the design evaluation is the fissile material data interpretation.

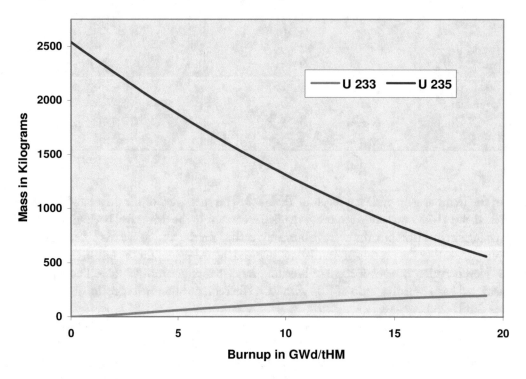

Figure 6.6: Plot of uranium isotopes versus burn-up for the driver fuel.

Figure 6.6 implies that U-233 production is a continuously increasing and there is no benefit to lowering the fuel burn-up of the thermal breeder for maximizing U-233 production. The data has been plotted beyond the attained burn-up to show the non-drooping trend of U-233 before average burn-up is reached. But for Pu-239 (as shown in Figure 6.7), the peak is attained close to the average burn-up of 18.72 GWd/tHM. The increasing trend of Pu-240 ensures the plutonium will be reactor-grade in the equilibrium state. At the end of the desired average burn-up of the driver and blanket fuel, the core comprises of 1445 kilograms of fissile material (U-235 = 667 kilograms, U-233 = 229 kilograms, Pu-239 = 303 kilograms & Pu-241 = 45 kilograms). The plutonium produced has a Pu-240/Pu-239 ratio of 50.5% in the driver fuel. U-233 and U-235 are a small percentage of total uranium and are mixed with plutonium, minor actinides and fission products, making the removal process uneconomical and making the fuel proliferation resistant.

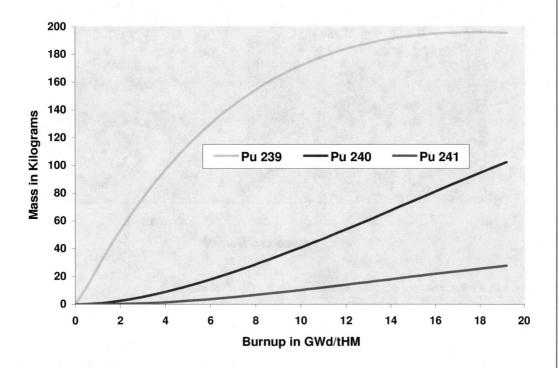

Figure 6.7: Plot of plutonium isotopes versus burn-up for the driver fuel.

Figures 6.8 & 6.9 depict the uranium and plutonium isotopes for the fuel configuration irradiated in the 152 radial blanket assemblies similar in shape and size to the driver fuel assemblies. The Pu-240/Pu-239 ratio reaches 72.8% in the radial blanket fuel. EOC uranium composition for the equilibrium stage is a 2:1 mixture of U-233 and U-235.

At the start of the first cycle, the radial blanket assemblies produced 2.7% (Table 6.4) of the total core heat and they constituted 1/3 of the total fuel assembly channels. This is because the fissile content in the blanket at the beginning of cycle is nominal. The contribution of blanket heat slowly increases with burn-up as production of fissile material rises. At the end-of-life (585 days) in the first cycle, the blanket heat is 7.7% of the total core heat generation. The blanket heat increases over the following cycles and finally reaches 15.6% for the equilibrium cycle.

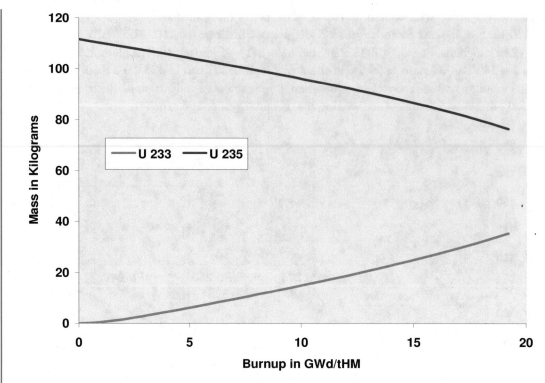

Figure 6.8: Plot of uranium isotopes versus burn-up for the blanket fuel.

As observed, there is an asymmetry in the power produced between driver and blanket fuel assemblies. To avoid large coolant temperature asymmetries, proper flow distribution among channels must be maintained. In PWR-type thermal breeder reactors, the temperature variability is evened out by cross flow; but this depletes the coolant from blanket assemblies and degrades the cooling of blanket fuel pins. A better solution is to surround the PWR-type assemblies with can walls, segregating the sub-assemblies. In CANDU reactors, this segregation is inherent. The flow in individual channels is controlled with orifices and ventures. Coolant mixing of selective channels at the core outlet maintains the uniform temperature distribution across the channels and core as a whole.

As stated in Table 6.4, there is a reduction in the total power production for TBR-2 compared to the CANDU-6 core. The TBR-2 core uses more fissile content than CANDU-6 because of huge parasitic absorption by fertile materials that are not efficient fissile material producers. Thus, the intermixing of Th-232 with the uranium in the driver fuel is not only required from a non-proliferation point of view but also for an efficient fuel cycle.

The rate of reactivity loss per day of TBR-1 core is close to that of CANDU-6 fuel, implying that this core can be operated with similar regulation and protection control systems. Near equal

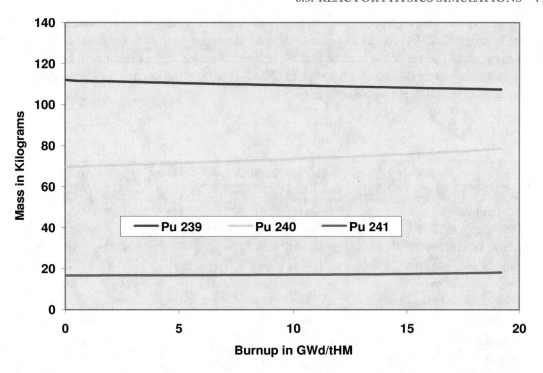

Figure 6.9: Plot of plutonium isotopes versus burn-up for the blanket fuel.

Table 6.4: Power in Zones for the First Cycle of TBR-1			
	TBR - 1	TBR - 2	CANDU - 6
Core	3214 MW$_{th}$	2427 MW$_{th}$	2552 MW$_{th}$
Driver	3127 MW$_{th}$	2425.4 MW$_{th}$	–NA–
Blanket	87 MW$_{th}$	1.6 MW$_{th}$	–NA–
Percentage of Heat Generated in the Blanket	2.7% - BOC 7.7% - EOC	0.065% - BOC	–NA–

initial multiplication factor for the TBR-1 and CANDU-6 fresh fuel core makes the reshuffling sequence similar during the pre-refueling phase. The online refueling sequence to maintain excess reactivity can be similar too.

6.3.4 OPTIMIZED DESIGN
In total, 304 driver fuel assemblies are surrounded by 152 radial blanket assemblies in the optimized design with the assumed constraints. No non-uniformity is needed in the lattice pitch dimensions,

fuel geometry and axial arrangement of fuel bundles. The power density of the thermal breeder fuel is similar to the CANDU fuel at 32 watts of thermal power per gram of heavy metal so as not to alter the pumping power of the coolant for heat removal. Because of the similarity in lattice pitch dimensions and power density of TBR with CANDU-type core, the quench time and reflooding sequence during LOCA are assumed to be similar.

Table 6.5: Dimensions, Initial Reactivity and Refueling Sequence		
Descriptions	TBR - 1	CANDU-6
Fuel Pin Radius	0.6077 cms	0.6077 cm
Fuel Clad Gap Thickness	0.0043 cms	0.0043 cm
Clad Thickness	0.042 cms	0.042 cm
Pincells in a Bundle	37	37
Fuel Bundle Length	50 cms	50 cm
Pressure Tube Inner Radius	5.1889 cms	5.1889 cm
Pressure Tube Outer Radius	5.6032 cms	5.6032 cm
Calandria Tube Inner Radius	6.4478 cms	6.4478 cm
Calandria Tube Outer Radius	6.5875 cms	6.5875 cm
Assembly Channel Pitch	28.575 cms	28.575 cm
Each Assembly Channel has	12 Fuel Bundles	12 Fuel Bundles
Number of Driver Fuel Assembly Channels	304	320
Number of Blanket Assembly Channels	152	–NA–
Number of Fuel Reshuffling Days	533 (first cycle)	628
Average Burn Up (GWD/MTHM)	18.72 (first cycle)	22.08
Days to Reach Average Burn-up	585 (first cycle) 306 (equilibrium cycle)	690
Initial K-effective	1.47921 (first cycle)	1.45922
Power Density in Watts/Grams	32	32
Coolant & Moderator	D_2O	D_2O
Reactivity Loss per Day	0.803 milli-k (highest of all cycles)	0.649 milli-k

D_2O at low temperature and pressure is used as the reflector to surround the radial blanket assemblies. There is no physical boundary between the reflector D_2O and the moderator. The moderator cooling and purification system can be designed to accommodate the extra D_2O volume from the reflector. Larger pitch length, similar to that of a CANDU-6 core improves the moderator-to-fuel-atom ratio compared to fast breeder reactors. The neutron leakage from the driver fuel assemblies has been recovered by radial blanket assemblies and the surrounding reflector, which improves the production of fissile material. An important process following the fissile material production is spent fuel reprocessing. Handling of spent thorium fuel is complicated by the radiological hazards of Th-228.

Th-228 is generated by the following decay and burn-up chain.

$$ {}^{232}_{90}Th \xrightarrow[(n,2n)]{} {}^{231}_{90}Th \xrightarrow[(\beta^{-1})]{} {}^{231}_{91}Pa \xrightarrow[\sigma(n)]{} {}^{232}_{91}Pa \xrightarrow[(\beta^{-1})]{} {}^{232}_{92}U \xrightarrow[(2\alpha^4)]{} {}^{228}_{90}Th $$

$$ {}^{232}_{90}Th \xrightarrow[\sigma(n)]{} {}^{233}_{90}Th \xrightarrow[(\beta^{-1})]{} {}^{233}_{91}Pa \xrightarrow[(\beta^{-1})]{} {}^{233}_{92}U \xrightarrow[(n,2n)]{} {}^{232}_{92}U \xrightarrow[(2\alpha^4)]{} {}^{228}_{90}Th $$

Thorium-228 decays into high energy gamma emitters, but its accumulation can be reduced by neutron capture of its predecessor U-232. The two paths are shown below.

$$ {}^{228}_{90}Th \xrightarrow{(\lambda)} \underset{(\gamma emitter)}{{}^{212}_{83}Bi} \xrightarrow[(2\alpha^4)]{} \underset{(\gamma emitter)}{{}^{208}_{81}Tl} $$

$$ {}^{232}_{92}U \xrightarrow[(\sigma(n))]{} {}^{233}_{92}U $$

The decay to Bi-212 and Tl-208 nuclides is the major radiation problem because the daughter nuclides emit highly penetrating 2.6 MeV gamma rays, making personnel shielding difficult or impractical. In a fast breeder, the fissile material density is large enough to provide high neutron flux that would enhance neutron capture in U-232, but fast neutrons also enhance n, 2n reactions in Th-232 and U-233, which are the source of the Th-228 in the first place. In thermal breeders, the large moderator to fuel volume ratio and small fuel rod radius effectively reduce the (n, 2n) reactions of Th-232 decreasing production of U-232. The modeled TBR-1 produces 1.51 grams of Th-228 in the driver fuel and only 0.0266 grams in the blanket fuel. This fuel configuration would not raise the radiation hazard levels in reprocessing compared to that of the existing spent fuel from thermal uranium reactors. The TBR-1 design with the optimized fuel configuration is economically beneficial and proliferation resistant because of its higher η value (as shown in Table 6.6), reactor-grade plutonium isotope mixture, low fissile content uranium isotopes, high burn-up and on-power refueling sequence similar to the existing CANDU reactors.

Figure 6.10 shows the flux at the beginning and end-of-life in the first cycle for the TBR-1 core. The flux is larger when the entire fuel assembly is considered as the control volume instead of a single fuel rod because of increased moderator to fuel atom ratio.

As stated earlier, a higher moderator to fuel ratio improves upon the 'η' by decreasing 'α'. Any on-power refueling sequence or reshuffling of the driver fuel for distributing the heat generation

Table 6.6: Beginning of Life α and η Coefficients			
	TBR - 1	TBR - 2	CANDU - 6
α for Driver	0.067067	0.067434	0.0675844
α for Blanket	0.050618	Blanket does not have fissile content	Blanket does not exist
α for the Core	0.064526	0.067434	0.0675844
$\eta = \frac{\nu}{1+\alpha}$	2.309	2.295	2.295

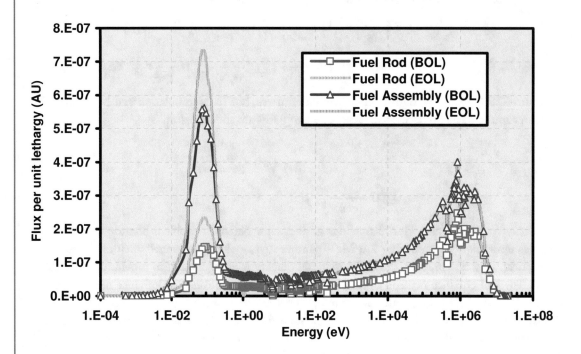

Figure 6.10: Beginning and end of cycle flux pattern for TBR-1.

should not lead to a shift in the energy range. If the flux increases around 0.1 eV of neutron energy, not only 'η' drops but the captures cross-sections of xenon and samarium reach the peak values increasing the parasitic absorption. Like any other reactor system, TBR righteously depends on the fuel configuration, fuel mapping in the core and heat distributions to achieve a breeding ratio greater than 1 by not losing criticality and controllability (large changes in reactivity in a unit time).

6.4   IMPLEMENTATION OF ALTERNATE REACTOR SYSTEM IN INDIA'S NUCLEAR FUEL CYCLE

The U.S.–India nuclear accord requires a nuclear power program with international collaboration on fuel verification and monitoring. This program is supposed to provide the effective use of resources while addressing proliferation concerns as an alternative to the presently pursued three-stage-power-program. The proposed alternate nuclear reactor system is based on the following objectives:

- Optimization of electricity production,

- Minimization of international uranium use,

- Maximizing proliferation resistance and

- Minimization of domestic uranium use.

   A comparison between the alternate fuel cycle and the present three-stage-program is presented in this section. It was observed that the nuclear power program that better meets the objectives by the year 2030 would serve the long term energy needs of India. International collaboration can significantly assist in meeting the energy needs of India, and also simplifies the objectives of the alternate reactor system.

6.4.1   DESCRIPTION OF ALTERNATE FUTURE FUEL CYCLE

The suggested model nuclear power program, with thermal breeders as the next stage after PHWRs, can appropriately fit the nuclear fuel cycle of nations like India having CANDU spent fuel along with thorium reserves. The handling of end of cycle fissile content of these TBRs is less of a challenge than in higher burn-up thorium reactors or fast reactors because of the reduced presence of Th-228 and its hard gamma emitting daughter nuclides in the bred fuel. Initially, the driver fuel for this core is made from internationally supplied low enriched uranium, and by default comes under a monitoring and verification process. The TBR reactor system under these safeguards and verifications will require less enriched uranium over time, eventually being fueled by 20% un-enriched (natural) uranium and 80% domestic thorium reserves, assuring life time fuel supplies. The mixing of thorium into the driver fuel has been shown in the previous sections to provide proliferation-resistant EOC spent fuel, because the fissile uranium isotopes U-233 and U-235 occur at low enrichment concentrations relative to the U-238, and the plutonium is produced with a reactor-grade ratio of Pu-240/Pu-239. The thorium-uranium mixed driver fuel was also found to be economical in producing sufficiently high burn-up, using existing CANDU reactor geometry, and using India's thorium resources. This alternate reactor system (shown in Figure 6.11) includes all the necessities of a fuel cycle for a nation seeking breeder systems to exploit the combination of thorium reserves and tonnes of spent fuel from thermal reactors. Because the fresh core driver fuel does not use plutonium, the system avoids the international concerns of proliferation that would otherwise be associated with plutonium extraction and transfer from the spent PHWR fuel to the TBR stage of the program. Later stages of spent fuel recycling are included in the monitored and safeguarded phase of the TBR fuel cycle and are therefore conducive to international cooperation.

The proposed nuclear power production strategy shown in Figure 6.11 can replace or operate in parallel with the presently pursued three-stage-program to meet the energy demands of the country. This alternate reactor system is a step forward toward international safeguards of reprocessing facilities. There can be fuel handling and reprocessing facilities explicitly dedicated to the proposed alternate reactor systems. The two most positive aspects in this power program are well accounted low enriched uranium from international suppliers and easily verifiable thorium supply for the breeder cycle, which makes long-term international cooperation feasible and sustainable.

6.4.2   METRICS OF INTEREST

Developing a fuel composition in a given geometry (CANDU-type in this study) with optimized distribution of driver and blanket assemblies was the first step in realizing the metrics of objectives for the modeled alternate reactor system. The basis of these metrics was described at the beginning of this Section 6.4. The first row of the metrics in Table 6.7 shows the total uranium utilized per unit of heat generated. As can be seen, the blanket fuel consumes less uranium per unit of energy delivered than the driver fuel. This is because it has lower fissile uranium content and a significant contribution of fission heat from plutonium and minor actinides. Thus, the blanket performs almost like a waste transmuter.

The blanket fuel has high Pu-240/Pu-239 ratio, and irradiation in the core increases this ratio, making the plutonium less suitable for weapons proliferation. Table 6.7 also shows that the fissile uranium content remains far below the weapons-grade limit of 20% concentration.

Table 6.7: Metrics for Assessment of Proposed Nuclear Power Program				
Metrics	Driver Fuel		Blanket Fuel	
of Objectives	BOL	EOL	BOL	EOL
Total U / GWd	–NA–	17.84 kilograms/GWd	–NA–	11.5 kilograms/GWd
Pu-240/Pu-239	0%	50.5%	62.5%	72.8%
Total Pu	0 kilograms	529 kilograms	208 kilograms	215 kilograms
(U-233+U-235) / U	4.25%	1.31%	0.31%	0.31%
Pu-240	0 kilograms	99 kilograms	70 kilograms	78 kilograms

6.4.3   COMPARISONS TO EXISTING INDIAN FUEL CYCLE

The energy output of this thermal breeder reactor system was calculated with the assumption of 30% overall plant cycle efficiency and 75% operational capacity factor for each power plant. As modeled

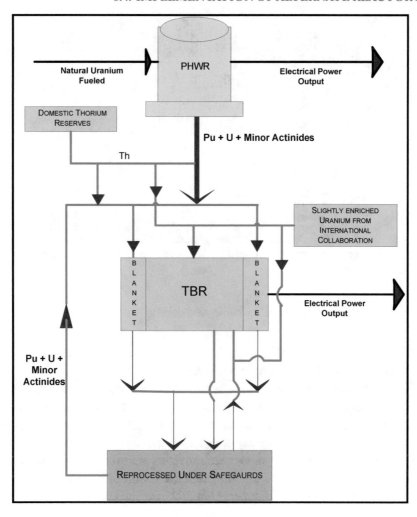

Figure 6.11: Proposed alternative nuclear power production strategy for India.

in SCALE5.1, the input deck for a TBR-I core has thermal output 3214 MWth. This thermal heat with the above assumptions would produce approximately 725 MWe (time-averaged).

Compared with present PHWRs and the PFBR now under construction, the proposed TBR program output grows twice as fast (500 MWe PHWR at 75% capacity factor = 375 MWe). Because of the similarity in design to a CANDU-6 core, the TBR power plant would have coincident secondary and tertiary loops. The proposed TBR power projects could be developed in the same time frame as planned PHWR projects. Representing these projects on the basis of the metric of total

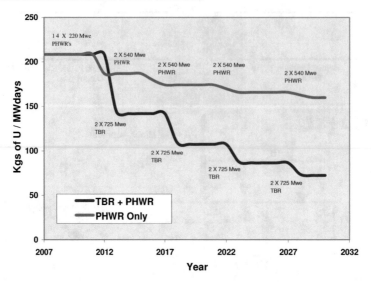

Figure 6.12: Comparison for uranium utilization in electricity generation.

uranium/GWd, Figure 6.12 suggests that there is a substantial savings in the quantity of uranium required by the safeguarded TBR reactors per unit energy delivered.

Starting with fourteen PHWRs, all the future reactors would be under safeguards. If the currently prevailing plan were replaced with a TBR power program beginning in 2012 and extending beyond 2030, there would be a substantial advantage in electricity production per unit of uranium. Under the U.S.–India civilian nuclear cooperation agreement, India would have to supply fuel to eight PHWRs; without the agreement, the domestic uranium reserves are required for all 22 reactors along with future PHWRs. Figure 6.13 shows that the break even point of lowering the use of domestic uranium per unit of electricity production occurs in 2021 for the currently planned construction of PHWRs. Figure 6.12 shows that this break-even point could occur as early as 2012 with the proposed TBR power program.

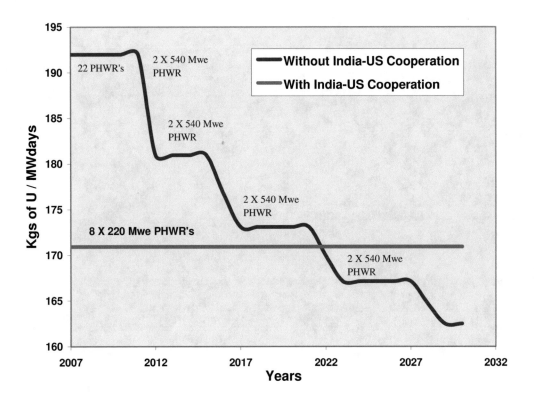

Figure 6.13: Comparison of domestic uranium use with/without cooperation.

CHAPTER 7

Conclusions

As an ambitious power-seeking state, India has, to a large extent, mastered the components of nuclear fuel cycle, with far-reaching consequences in both civilian and military sectors. It started as a strong proponent of nonproliferation of nuclear technology, but ended up as nonsignatory to the NPT, on grounds of disparity. On many past occasions efforts have been made to bring the Indian nuclear sector under safeguards. The current U.S.–India nuclear cooperation agreement appears the closest yet to a credible and possibly sustainable non-proliferation initiative.

With this monograph, an effort has been made to study all the components of the nuclear fuel cycle and assess the detailed implications of international cooperation on the weapon and energy programs. The weapon program does not seem to be deterred or assisted by the U.S.–India civilian nuclear cooperation, as the present rate of plutonium production can be sustained with the available uranium reserves and production reactors, and all additional materials made available to India under the agreement will be under international safeguards and therefore unavailable for weapons usage.

This agreement makes fewer reactors available for weapons-grade plutonium production. In that respect, it definitely decreases the latency capabilities for production of weapons-grade plutonium from that of the scenario of failed agreement. Furthermore, the reactor constraint of only 8 unsafeguarded reactors decreases the reactor-grade plutonium production.

On the energy front, India would have to operate its commercial PHWRs at a lower capacity factor without the agreement. This phenomenon would increase the quantity of lower-grade plutonium produced from the present levels. Further increase in plutonium production by lower capacity factor of PHWRs or a newer facility leads to excessive accumulation and a concern for Southeast Asia arms race.

A small but significant component of India's fuel cycle is enrichment facility. The need for enriched uranium is centered about its use in nuclear submarine cores, thermonuclear devices and reactor fuel. The constraints to enrichment are not the natural uranium feed but the technological hurdles. The requirement for nuclear fuel for submarines is not affected by the U.S.–India cooperation agreement because of the meager amount of uranium required. Rather, technical expertise in modeling and building are the limiting factors.

Breeders are the route that India sees to the third stage of the nuclear power program, and therefore the essential building block for an ultimate vibrant energy program. The third-stage reactors are to be fuelled with U-233 obtained from the thorium breeding in the fast breeder reactors of second-stage. The route to this third stage likely would be easier if India could use plutonium derived from international uranium to fuel some of its fast breeders. However, this is likely to be a sensitive issue, because of the perceived potential to use this plutonium for weapons purposes.

Through this book an alternative to fast breeders has been suggested. A power program with PHWRs followed by thermal breeder reactors constrains proliferation risks, lowers uranium use and establishes a thorium breeding cycle. Unlike the FBR, the thermal breeder modeled in Chapter 6 makes reactor-grade plutonium, does not impart a huge radiation risk from spent fuel because of lower quantities of Th-228, uses existing core geometry and does not involve a sodium-water heat transfer phase.

The suggested thermal breeder design appropriately fits to the nuclear fuel cycle of states having PHWR reactor experience and thorium reserves. The EOC fissile content is easy to recover because of negligible production of Th-228. The fuel configuration is proliferation resistant because of lower ($< 20\%$) percentage of U-233 and U-235 in the fuel and reactor-grade nature of plutonium. The reactor can be brought under safeguards and international monitoring system by involving the international supply of low enriched uranium. Presence of thorium fuel fits the necessities of the fuel cycle that are looking for breeder systems for exploiting the domestic reserves of thorium. By not incorporating any plutonium in the driver fuel, this system bypasses the need for reprocessing of thermal reactor spent fuel for extraction of plutonium.

In the long run, the TBR cycle appears to be better positioned with regards to the quantity of uranium used per unit of electricity generated. This cycle can be implemented in a manner that partially enhances the burn-up of the EOC fissile content of the PHWR spent fuel and better attains a breeder equilibrium state.

Through this study a feasible nuclear power program has been suggested as an alternative to that of the presently pursued via the three-stage power program. A viable international nuclear collaboration can be established on the basis of safeguards and verification methods for the proposed alternate nuclear fuel cycle. The present, closely guarded fuel cycle of India has large scope for improvement by integration with the international domain. The objective of separation of strategic and civilian nuclear sectors can be more explicitly achieved with the proposed thermal breeder nuclear power program.

CHAPTER 8

Acknowledgements

This monograph is a significant revision and extension of a 2007 dissertation[1] submitted by Dr. Taraknath V.K. Woddi, to Texas A&M University, in partial fulfillment of the doctoral requirements in nuclear engineering. Dr. William S. Charlton served as Chair of the Advisory Committee for the research that underpinned that work. Dr. Paul Nelson was a member of that Advisory Committee. Dr. Nelson contributed to Chapters 1–5 and 7 of this monograph.

A book aspiring to be a comprehensive brief of the India's Nuclear Fuel Cycle in the era of U.S.–India nuclear cooperation could not have been written without detailed opinions from a number of experts. There are many aspects to a nuclear fuel cycle, other than creating a reverse model and extracting thereby the material accounting details. This work benefitted from critical scrutiny by a number of very able individuals having a wide range of experience in nuclear history and its relevance. The insights and research by eminent academicians dedicated to the understanding of complex intermingled weapon and civilian nuclear fuel cycle have lead to the making of this monograph.

The authors especially would like to thank Dr. Marvin L. Adams, whose intellectual standards provide a constant role model for aspiring nuclear engineers. As and when time permitted, he provided a wealth of information to focus this study with his well-researched knowledge base. Much of the information analyzed herein through nuclear reactor theory has its roots in his astute approach to teaching.

We would want to thank Dr. Jean Ragusa for all the support and help. Without his encouragement, suggestions, and devotion to the topic this book might not have been written. He is a teacher with in-depth subject knowledge and has shown profound interest in the study. We would like to express our gratitude to Dr. Ken Ricci for editing and proofreading the manuscript with particular emphasis on his favorite baby "Thermal Breeder Reactor." We are greatly indebted to fellow Aggies David Ames, Ayodeji and others for their indispensable advice and help in matters of core analysis and approach to system modeling. Without their generosity this book would be sorely lacking impact.

Taraknath K.V. Woddi extends warm thanks for the unrelenting support of Scientech and fellow employees Jim Chapman, Jeffrey Julius, Bob Bertucio, Lincoln Sarmanium and others in encouraging the production of this manuscript. Scientech, and Curtiss-Wright its parent company, along with Texas A&M University and its Nuclear Security Science and Policy Institute, long realized

---

[1] Taraknath Woddi Venkat Krishna, Nuclear Fuel Cycle Assessment of India: A Technical Study for U.S.–India Cooperation, Ph.D. dissertation, Texas A&M University, December 2007, URL http://repository.tamu.edu/bitstream/handle/1969.1/85860/Woddi.pdf?sequence=1, accessed July 5, 2009.

the importance of US-India accord and furthered nuclear technology as a viable economic pursuit. This book stands as a token of the foresight of these institutions.

I, Taraknath Woddi, thank my loving wife Divya and our bright future Siddharth, for whose generation presumably the cooperation agreement matters the most. Many times Divya had to bear with my erratic time schedule while I was involved in the research. Without her support and love, I would not have been able to march a single step ahead. Ton of thanks to my parents and in-laws for listening uncomplainingly to every piglet story and did not lose patience even when I was in challenging times.

Paul Nelson is deeply appreciative of the patience and wonderful sense of humor his wife, Diana, has faithfully maintained over years of many such seemingly endless projects.

This material is based upon work partially supported by the Department of Energy, National Nuclear Security Administration, under Grant DE-FG52-06NA27606, titled "Support for the Nuclear Security Science and Policy Institute, Texas Engineering Experiment Station at Texas A&M University." Any opinions, findings, and conclusions or recommendations expressed in this publication are those of the authors and do not necessarily reflect the views of the National Nuclear Security Administration or the Department of Energy.

Biography

TARAKNATH V.K. WODDI

Dr. Woddi serves as the Risk and Reliability Analyst for Scientech: A Curtiss–Wright Flow Control Company and provides consulting services to nuclear power plants for meeting the regulatory standards. Dr. Woddi served as licensed reactor operator at CANDU nuclear power plant and has vast experience with plant systems operations and troubleshooting. Dr. Woddi has core/fuel design expertise through thermal breeder reactor modeling initiative. He wrote the first thesis on the RACE project, an endeavor of waste transmutation in May 2005. He widely contributed to non-proliferation studies through assessment of nuclear fuel of India. He is presently in the industry of nuclear power plant safety and risk assessment. He has published articles on the feasibility of temperature feedback accelerator-driven sub-critical reactor systems, nuclear fuel cycle of India and impact of the U.S.-India nuclear accord. Dr. Woddi earned his Master's and Ph.D. in Nuclear Engineering from Nuclear Engineering Department of the Texas A&M University.

WILLIAM S. CHARLTON

Dr. Charlton serves as the Director of the Nuclear Security Science and Policy Institute (NSSPI) at Texas A&M University (TAMU) and as an Associate Professor in the Nuclear Engineering Department at TAMU. NSSPI is a multi-disciplinary organization that coordinates research and education programs in the area of nuclear nonproliferation and nuclear material safeguards. NSSPI customers include Los Alamos, Sandia, Oak Ridge, and Lawrence Livermore National Laboratories as well as the National Science Foundation (NSF), Office of Defense Nuclear Nonproliferation (DOE/NA-20), Office of Nuclear Energy (DOE/NE), Defense Intelligence Agency (DIA), Domestic Nuclear Detection Office (DNDO), and the International Atomic Energy Agency (IAEA). As Director of NSSPI, Dr. Charlton directs the overall NSSPI activities and personnel. Prior to his appointment at TAMU, he was an Assistant Professor in the Nuclear and Radiation Engineering Program at the University of Texas at Austin from 2000-2003. From 1998-2000, Dr. Charlton was a Technical Staff Member in the Nonproliferation and International Security Division at Los Alamos National Laboratory (LANL). He teaches courses at TAMU which study the technical aspects of nuclear nonproliferation, nuclear material safeguards, and international nuclear security as well as fundamentals of nuclear reactors and nuclear fuel cycle systems. Dr. Charlton earned a Ph.D. in Nuclear Engineering from Texas A&M University. Among his many awards, Dr. Charlton was named the George Armistead Jr. '23 Faculty Fellow at TAMU in 2005. Dr. Charlton is recognized as one of the leaders in nuclear nonproliferation education and research and he has over 140 technical publications in refereed journals and conference proceedings.

PAUL NELSON

Paul Nelson is Professor Emeritus of Computer Science, Nuclear Engineering and Mathematics, at Texas A&M University, where he currently serves as Associate Director for International Programs in the Nuclear Security Science and Policy Institute. He is a Fellow of the American Nuclear Society. Professor Nelson has previously held positions with the Oak Ridge National Laboratory, Sandia National Laboratories, and Texas Tech University. Additionally he has held visiting positions at the Georgia Institute of Technology and the California Institute of Technology.

Printed in the United States
by Baker & Taylor Publisher Services